Agricultural Communications

Changes and Challenges

Agricultural Communications

Changes and Challenges

Kristina Boone

Terry Meisenbach

Mark Tucker

Kristina Boone received her doctorate in Extension education from Ohio State University. As an agricultural communicator, she has worked for a daily newspaper, in a public relations firm and in Cooperative Extension Service. Dr. Boone presently is coordinator of the Agricultural Journalism Program at Kansas State University in Manhattan, where her major responsibilities include teaching and research.

Terry Meisenbach is the director of Communications and Information Access at USDA-CSREES/CTDE in Washington, D.C. His professional experience includes 18 years as an agricultural communications professor and publications coordinator at the University of Nebraska–Lincoln, where he is completing a doctorate in vocational and adult education, with emphasis on distance learning and minority audiences.

Mark Tucker received his doctorate in rural sociology from Ohio State University. He has taught courses in agricultural communications and agricultural journalism at Texas Tech University and the University of Missouri–Columbia, and has served as an agricultural publications editor for Ohio State University Extension. Dr. Tucker presently is an assistant professor of agricultural communications at Ohio State University, where his major responsibilities include teaching and research.

Iowa State Press
A Blackwell Publishing Company
2121 State Avenue, Ames, Iowa 50014

Orders: 1-800-862-6657
Office: 1-515-292-0140
Fax: 1-515-292-3348
Web site: www.iowastatepress.com

First edition, 2000
First paperback edition, 2003

Library of Congress Cataloging-in-Publication Data

Boone, Kristina
Agricultural communications : changes and challenges / Kristina Boone, Terry Meisenbach, Mark Tucker.—1st ed.
p. cm.
Includes bibliographical references (p.).
ISBN 0-8138-2157-6 (hardcover); 0-8138-2167-3 (paperback)
1. Communication in agriculture. I. Meisenbach, Terry. II. Tucker, Mark A. III. Title.
S494.5.C6B56 2000
630'.1'4—DC21 00-023757

Text photographs courtesy of Photographic Services, Kansas State University.
Front cover: Apollo 17 lunar mission photograph, courtesy of the National Space Sciences Data Center and Dr. Frederick J. Doyle, Principal Investigator.

The last digit is the print number: 9 8 7 6 5 4 3 2 1

Contents

Preface

The idea for this textbook was born out of several years of conversations about the difficulty of finding suitable teaching materials for use in agricultural communications and agricultural journalism. During this time, we also began to note with interest, and curiosity, the different administrative and professional philosophies at work in building academic agricultural communications and agricultural journalism programs nationally. While these different approaches were based partially on the unique needs of various colleges and universities, they also reflected the widely different worldviews of university administrators, faculty and practitioners about the mission and objectives of these programs.

The practical need for teaching materials, coupled with concerns about the academic underpinning of the discipline, convinced us of the need to respond to both issues. The result is a text that addresses a number of historical milestones as well as critical issues that will shape the future of agricultural communications, both as a profession and as an academic discipline. The primary audience for this book is agricultural communications students. Secondary audiences include agricultural communications practitioners and scholars, both of whom have a stake in the topics and issues we address.

The purpose of this book is to provide readers with a snapshot of agricultural communications at the beginning of the 21st century as well as fodder for discussion regarding its future. The title, "Agricultural Communications: Changes and Challenges," emphasizes the transitional state of the field. Currently we are maneuvering through the information age—discussing its promises and challenges and using its products. Certainly the information age was spurred by technology; however, the age of technology is extending beyond the age of information. This book strives to present the impact of the age of information on agricultural communications.

The book focuses on the ages traversed by agricultural communications and acknowledges that it is still in transition. It not only provides background regarding the profession and its literature, but also information on agriculture in general, because of the integral tie between the profession and the food and fiber industry. The book also looks toward the future.

Two additional elements that may help spur critical thinking among students are the Professional Perspectives and Nexus Points. We requested contributions from several professionals in agricultural communications regarding the field and where it is heading. Providing these insights were Jeff Altheide of Gibbs & Soell Communications; James Evans and Robert Hays, professors emeritus from the University of Illinois; Donna French Dunn of the American Agricultural Economics Association; Scott Kilman of *The Wall Street Journal;* Loren Kruse of *Successful Farming;* and Ricky Telg of the University of Florida. The Nexus Points provide issues that can be used for discussion about our field. They appear throughout the book. Chapter 5 presents our perspectives on these points.

It is unlikely that all stakeholders will agree on the essential elements of a textbook in agricultural communications. We view this first edition as a work in progress. Feedback from agricultural communications students, practitioners, teachers and researchers will be necessary to refine its content in order to serve best the multiple uses for which it is intended.

At the same time, we share the view that many of the issues facing the discipline are open to controversy and debate among various groups. While it is apparent that academic and professional interests do not always coincide on the state and future of the discipline, it is also true that academicians and practitioners are frequently in conflict among themselves about these same issues.

Such tensions are common in all disciplines, and they need not be divisive. Indeed, the educational approach adopted in this textbook views differences of opinion as entrance points from which students and professionals can become engaged in healthy debate about the mission and future of the discipline. Students who study their craft from a critical perspective will be much better equipped to make judgments about their professional development than those who do not.

The collective outcome of these judgments will dictate the course of development for agricultural communications well into the 21st century.

Kristina Boone, Kansas State University
Terry Meisenbach, U.S. Department of Agriculture
Mark Tucker, The Ohio State University

Acknowledgments

We wish to thank several individuals for their contributions to this book. First, we wish to thank the professionals who contributed quotes: professors emeritus James Evans and Robert Hays, Scott Kilman, Jeff Altheide, Donna French Dunn, Loren Kruse and Ricky Telg. We benefited greatly from their insights and believe that our readers will as well. Second, we wish to thank Larry Whiting, Bob Furbee and Sherrie Whaley, who, over the years, have discussed with us many of the issues presented in this volume. Professors emeritus Delmar Hatesohl, Dick Lee and Claron Burnett have also influenced our thoughts about the profession. Finally, a special thanks is extended to our families, who have supported us throughout this endeavor.

Agricultural Communications

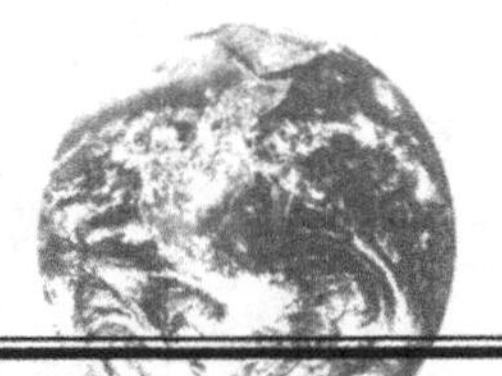

Changes and Challenges

1

Agricultural Communications Across the Ages

The first mediated communications about agriculture in the United States started during the early- to mid-19th century. Prior to that time, agricultural information was passed from farmer to farmer by word of mouth, and most of this information had come from Europe.

After the American Revolutionary War, advances were beginning to be made in agriculture, including attempts to develop technologies and practices appropriate for the colonies (Figure 1.1). Before 1780, like much of culture and business in the states, agriculture followed the patterns and practices established in Europe. Likewise, the first publications on agriculture that appeared in the United States were from Europe.

Several European agricultural books were published in English before the Revolutionary War, but the earliest U.S.-produced books were printed in the early 1800s. Two early volumes were the 1814 *Farmer's Assistant,* by John Nicholson, and the 1826–1827 *Farmer's Library,* by Leonard E. Lathrop.[1] The first American periodical, the *Agricultural Museum,* was published in 1811.[2] The U.S. Congress also began publishing agricultural information at this time, with its first technical publication in 1828 on the rearing of silk worms.[3]

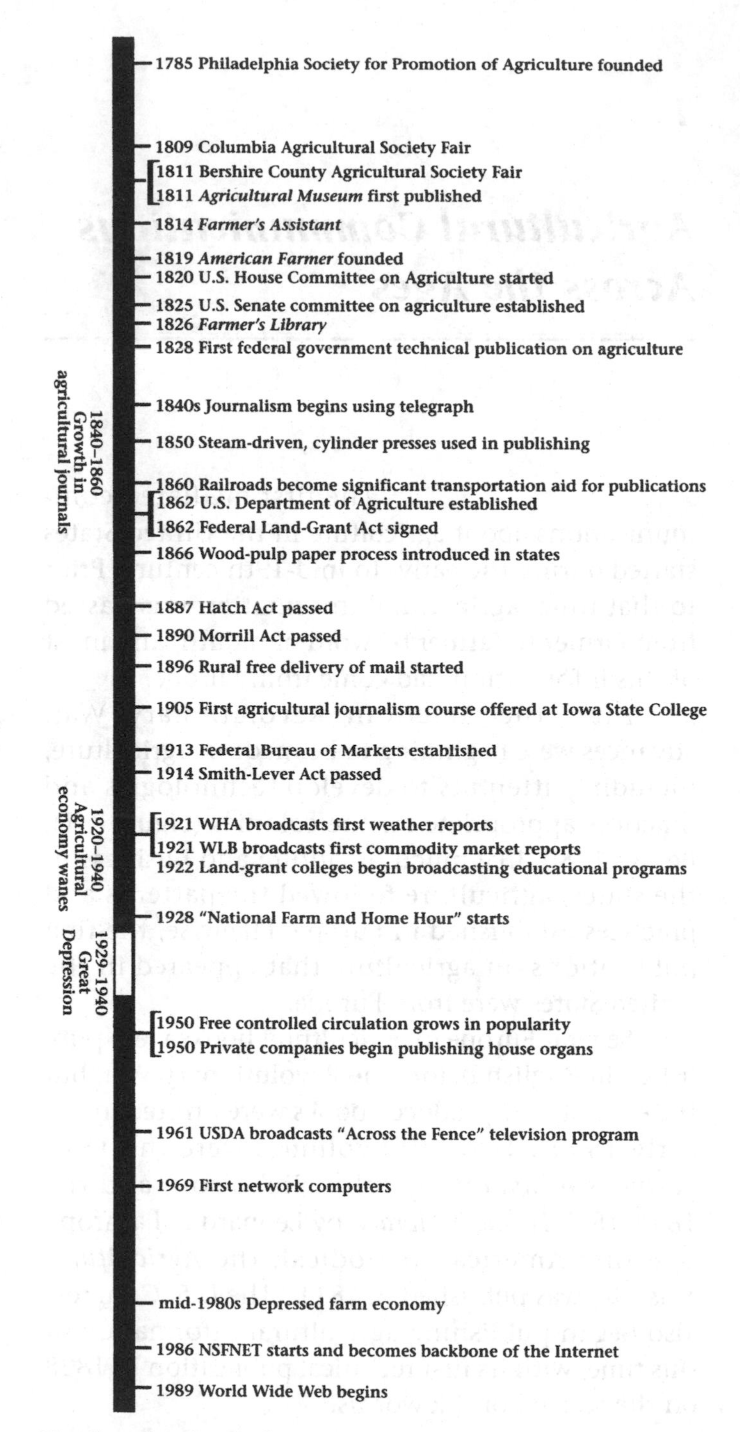

1785 Philadelphia Society for Promotion of Agriculture founded
1809 Columbia Agricultural Society Fair
1811 Bershire County Agricultural Society Fair
1811 Agricultural Museum first published
1814 Farmer's Assistant
1819 American Farmer founded
1820 U.S. House Committee on Agriculture started
1825 U.S. Senate committee on agriculture established
1826 Farmer's Library
1828 First federal government technical publication on agriculture
1840–1860 Growth in agricultural journals
1840s Journalism begins using telegraph
1850 Steam-driven, cylinder presses used in publishing
1860 Railroads become significant transportation aid for publications
1862 U.S. Department of Agriculture established
1862 Federal Land-Grant Act signed
1866 Wood-pulp paper process introduced in states
1887 Hatch Act passed
1890 Morrill Act passed
1896 Rural free delivery of mail started
1905 First agricultural journalism course offered at Iowa State College
1913 Federal Bureau of Markets established
1914 Smith-Lever Act passed
1920–1940 Agricultural economy wanes
1921 WHA broadcasts first weather reports
1921 WLB broadcasts first commodity market reports
1922 Land-grant colleges begin broadcasting educational programs
1928 "National Farm and Home Hour" starts
1929–1940 Great Depression
1950 Free controlled circulation grows in popularity
1950 Private companies begin publishing house organs
1961 USDA broadcasts "Across the Fence" television program
1969 First network computers
mid-1980s Depressed farm economy
1986 NSFNET starts and becomes backbone of the Internet
1989 World Wide Web begins

Timeline of major developments in American agricultural communications: 1785–1989.

> ***As an undergraduate student in agricultural journalism during the 1950s, I heard troubling reports about the future of agricultural publishing and broadcasting. Several national farm magazines with circulation in the millions had died, leaving observers puzzled and concerned. Some powerhouse radio stations, mainstays of early farm broadcasting in this country, were phasing out their farm programming. We heard predictions that farm broadcasting would be dead in five years as farm populations continued to drop in size. These and other concerns of the time did not dampen my professional interest. However, they remained with me as I moved into my career. And only through experience and historical analysis did I begin to understand more clearly the dynamics of agriculture-related journalism and communications. . . . Historical perspectives such as these have added greatly to my professional focus. They have helped me to see beyond the changes of the day and to put those changes into broader perspective.***
>
> ***—Jim Evans, professor emeritus, University of Illinois***

Following another trend started in Europe, in the late 1700s, agricultural societies in the United States began to form as agricultural knowledge became more specialized. In 1785 the Philadelphia Society for the Promotion of Agriculture was founded and included on its roll of members George Washington, Thomas Jefferson and Benjamin Franklin. Similar societies were founded in New York and Massachusetts in 1791 and 1792.[4] The greatest advantage of membership in these societies was access to their libraries, which collected books, magazines and newspapers and published listings of their holdings.

Societies provided excellent libraries, but wealthy planters also maintained private collections of documents. Both George Washington and Thomas Jefferson held significant collections of agricultural books. Many of Jefferson's books were donated later to the Library of Congress.

Agricultural societies began publishing in the 1790s. Their publications were devoted to disseminating practical information about farming. In general, these societies began their own publications out of frustration with the popular press, which used information supplied by agricultural societies only as filler and failed to present information that the societies considered substantial. Early society publications provided information on a diversity of topics, from the typical (cropping practices) to the more exotic (the use of elk as draft animals), and demonstrated a great interest in biological systems.[5]

Despite their efforts, at the end of the 18th century, agricultural societies were not diffusing agricultural knowledge as they had hoped. Readership and writers became more and more one and the same—large planters. After preaching to the choir for a while, many societies began trying different approaches in the early 1800s. Agricultural fairs were sponsored to promote exchanges of information. Two early examples of fairs were those of the Columbia Agricultural Society near Washington, D.C., in 1809, and the Berkshire County Agricultural Society in Pittsfield, Mass., in 1811.[6] These fairs were particularly useful in reaching "dirt farmers," the poorer farmers who typically had a disdain for "book farming" (using publications to guide their farm operations).[7] Dirt farmers preferred to see the innovations and talk about them. Thus, fairs were useful tools in reaching them. Fairs also became popular with other groups. Abraham Lincoln noted in an address the importance of fairs:

> Agricultural fairs are becoming an institution of the country; they are useful in more ways than one;

they bring us together, and thereby make us better acquainted, and better friends than we otherwise would be.... They render more pleasant, and more strong, and more durable, the bond of social and political union among us.[8]

The Massachusetts Agricultural Society hoped to create communications channels among farmers through publications, because the society maintained that each farmer had the knowledge to improve farming. The need, then, was to bring together these individuals' wisdom to form a cohesive body of knowledge. After years of failed attempts to gain contributions of articles from farmers, the *Massachusetts Agricultural Journal* halted publication in 1832.[9]

The First Journals

In the meantime, the first agricultural journals, the *American Farmer,* the *Plough Boy* and the *New England Farmer,* began publication. The content of these early publications, which were based in the East, included many society contributions as well as information on using manures, new crops and implements and improved livestock. The *American Farmer,* begun in 1819, was published by Baltimore postmaster John Stuart Skinner,[10] who is sometimes referred to as the Father of American Agricultural Journalism. Skinner was born on a Maryland farm and studied law before serving the federal government in various capacities. The *American Farmer* reflected Skinner's goals, which he explained in his first issue:

> The great aim, and chief pride of the *American Farmer,* will be to collect information from every source, on every branch of Husbandry, thus to enable the reader to study the various systems which experience has proved to be the best, under given circumstances.[11]

Such early agricultural editors had neither training nor practical experience in agriculture. Some, like Skinner and Solomon Southwick, who published the *Plough Boy,* were postmasters. Editors who began publications in the 1830s tended to develop more practical knowledge, although they still lacked an understanding of the scientific principles of farming. Making a profit from these early publications proved difficult, as was finding successors to carry on the publications. As a result, many of these publications were short-lived. Despite challenges, some thrived.

During the 1840s and 1850s, some agricultural publications grew large enough in popularity to become independent of the agricultural societies. As the magazines found independence and reduced their use of society material, state agricultural societies again began their own publications. The magazines then began stronger associations with U.S. colleges. College scientists began writing for these publications, even before land-grant colleges were established.

Monthly publications became especially popular in the 1850s. During that time, 80,000 copies of a single issue of the *American Agriculturist* were sold. The person responsible for making this magazine the best-selling farm journal of its time was Orange Judd. Judd, who had studied agricultural chemistry at Yale for three years with John Pitkin Norton, brought his scientific knowledge to the *American Agriculturist.* He was a critic of agricultural science for its lack of reliability and took on the role of protecting his readers from "innovations" that would not work. Norton, with whom Judd had studied, contributed articles to several journals based on his study of applying chemistry to agriculture. He held a chair of agricultural chemistry at Yale and, despite his death at age 30, made an impact by infusing agricultural practices with science. His work was carried on by Judd and Samuel W. Johnson.[14]

The U.S. government started to institutionalize interest in agriculture about this time. In 1820 the House of Representatives established the Committee on Agriculture. The Senate created a similar committee five years later.[12] However, most agricultural policy issues, such as tariffs, trade and transportation, were handled by other committees. The Senate disbanded its agriculture committee in 1857, but it reinstated the committee in 1862.[13]

Communications made a tremendous difference in the lives of rural families. In addition to business-related information, communications afforded rural families a view of the world.

Technology's Impact

In the latter half of the 19th century, technological innovations spurred growth in publishing, making printing cheaper and distribution more reliable. By the 1840s steam-driven, cylinder printing presses replaced labor-intensive, hand-powered, flatbed presses. Steam-driven, cylinder presses were 10 times faster than flatbed presses, but the new presses lacked quality and were inconvenient. However, these shortcomings were not an issue after stereotyping (casting plates for printing from molds) was perfected for curved plates in the 1850s. Another innovation, introduced in the United States in 1866, led to the development of a process to produce paper from wood pulp instead of rags, greatly reducing paper costs.[15] The U.S. news press also started using the telegraph in the 1840s, which helped speed news transmissions and resulted in standardizing the news story form (e.g., the inverted pyramid). The press used these innovations competitively and began employing individuals with technological skills.

Also during this time, metropolitan dailies began employing farm writers. Papers such as the *Chicago Tribune, The New York Times* and the *Des Moines Register* were addressing farm audiences, in part because they could finally reach them. Transportation aided technological advances in the press. In 1830 the United States had 23 miles of railroad track, but by 1860, 30,000 miles of track linked the country, allowing the news to reach rural America.[16] Later another innovation also helped in this effort. In 1896 free delivery of mail to rural areas was begun as an experiment, and by 1902, 5.8 million people were receiving free delivery, with more routes being added as quickly as possible.[17] With better delivery, farmers became an advertising market for newspapers. Agriculturally oriented publications also were boosted by the agrarian movement in the late 1800s and early 1900s. This movement produced the Populist Press, which covered some agricultural issues and reached a rural readership.

Federal Involvement

Another outside force had a tremendous impact on agricultural journalism. Justin S. Morrill, a U.S. congressman from Vermont, introduced the Federal

Land-Grant Act to Congress in 1857. Lacking support from key legislators, the bill failed initially but was passed in 1862. It granted 30,000 acres to each state for each senator and congressman representing that state. Each state was to sell the land and use the income to establish a college to serve the sons and daughters of farmers and mechanics and specialize in subjects related to those fields. Up until this time, higher education was restricted to the wealthiest Americans and was offered through private schools. Most states funded new colleges, whereas others funded existing ones. In 1890 Congress passed legislation to provide similar land allotments to historically black colleges of agriculture and regular funds to all land-grant colleges.

After the establishment of land-grant colleges, tensions arose over the goals of these institutions. In the 1870s and 1880s, members of farm organizations visited colleges and were displeased with what they found; they complained to legislators. The legislators were sensitive to the farmers, dismissing or forcing into resignation many faculty members, trustees and college presidents. College staffs battled back, and tensions continued to rise. Both sides wanted to produce a new type of farmer, but they disagreed about how to accomplish that goal. Farmers wanted to see the development of smarter farming techniques and wanted colleges to serve as agricultural business schools, teaching accounting, machinery handling and practical experience through daily manual labor. In addition, they believed the teachers should be successful, experienced farmers. College staffs, on the other hand, believed agricultural science to be critical to improving farming. While farmers could utilize scientific principles to improve agriculture, they could not produce those principles. Agricultural scientists, in their own estimation, were able to do so. Both groups battled until the 1890 Morrill Act satisfied both camps. In its provisions, the act required that land-grant institutions submit to federal oversight, making college records open to the public, which pleased the farmers. And because the act identified agricultural instruction as the study of the application of physical, natural and economic sciences to agriculture, the agricultural scientists were also pleased.[18]

At about this time, Congress, through the Organic Act of 1862, established the Department of Agriculture, following proposals from President Abraham Lincoln and the U.S. Agricultural Society, a private organization. However, government holdings of agricultural information started in the U.S. Patent Office, which established a separate agricultural division in 1839, and even distributed seed. The division later became the Department of Agriculture Library. By 1900, the library, later named the National Agricultural Library, was considered by many to house the most complete agricultural holdings in the world.[19]

Library holdings also became quite important as systematic research in agriculture came on the scene. Before 1880, little research was applied to agriculture in the United States. Due to the need for this type of study, Congress established agricultural experiment stations through the 1887 Hatch Act. Many state experiment stations, which were seen in Europe as early as 1834, started work prior to the Hatch Act. The first, the Connecticut Agricultural Experiment Station, was founded in 1875 and was not part of an agricultural college, although most that followed were. The stations began developing their own libraries and producing informational bulletins and volumes. By 1900 all states and territories had established experiment stations. Their libraries housed technical papers, whose audience was primarily other researchers. Other agricultural libraries provided broader offerings. Academic institutions began agricultural collections even before land-grant colleges were established. Land-grant colleges, however, helped establish a much greater number of agricultural libraries. These libraries were greatly underfunded for many years, but those in the eastern United States received significant contributions from state agricultural societies.

Another congressional act strived to move information from the colleges and researchers to the general public and farmers. The Smith-Lever Act of 1914 funded Cooperative Extension activities that had already begun at most agricultural colleges. To take advantage of the growing farm periodical base and promote the work of researchers and Extension specialists, agricultural colleges began hiring information specialists who edited publications and wrote research articles in a format acceptable to the public. Sometimes these communicators did little more than handle correspondence for the researchers, but their role began to grow. The farm press benefited from these editors because the content of their publications used more research-based information. Many of these communicators would later teach classes in agricultural writing. Iowa State College offered the first college course in agricultural journalism in 1905.[20]

Some academic programs are called agricultural journalism whereas others are called agricultural communications. Is there a difference? Outside of academia, is there a difference?

Nexus Point 1

Expansion

From the early days of agricultural publications, the farm press continued to expand steadily. With the opening of lands in the Midwest and West, more farms were seen in the central and western regions of the country. While total numbers of farm magazines and newspapers increased from 157 in 1880 to 400 by 1920, circulation numbers increased at an even greater rate, from about 1 million in 1880 to more than 17 million in 1920. Whereas in 1880 one in four farmers received a periodical, in 1920, the average farmer received two to three. The number of specialized publications also started to increase moderately.

Fueling the rise in circulations was the good general health of the economy, creating more subscribers and advertisers. Farmers with higher incomes were buying more implements and other labor-saving inputs, creating incentives for advertisers. The publications themselves were more accommodating to advertisers, whereas earlier periodicals considered advertisements to be foreign material. With more available income for investment, farmers were more interested in innovations. Farmers also were becoming less critical of "book farming" and were therefore more interested in the writings in the periodicals. And because of innovations in printing and the acceptance of advertising, magazine subscriptions were cheap. Whereas the earliest agricultural periodicals relied on state agricultural societies for information, the later ones had several sources, including land-grant colleges, state departments of agriculture, the U.S. Department of Agriculture (USDA), and other agricultural organizations, such as the National Grange and Farm Bureau.

By 1920 the editors of farm publications were changing their roles, as well. In the days of Judd and others, the editor was more of a subject matter authority, whereas the new editors were information movers. The more specialized a publication, though, the more educated the editor needed to be about the technical material. Publications also began addressing the entire rural family, with an emphasis on the fundamental value of rural life and agriculture. They used less personal journalism and more objective, third-person writing, except for columns, which featured case studies that used prescriptive statements to tell farmers what they should do. Printing advances also made design more appealing with better illustrations, photographs and even cartoons. Cartoon characters such as Peter Tumbledown in *Farm Journal,* Reckless Robert in *New England Homestead* and Lazy Farmer in *Prairie Farmer* were used for educational as well as entertainment purposes.[21]

Depression in the Farm Economy

The time between 1920 and 1940 was difficult for agriculture. Poor economic planning after World War I created a situation in which some sectors of the economy boomed (auto making, road and building construction) and others suffered (agriculture, coal mining, shipping). Farmers had increased production to meet wartime demands and were left overextended in the aftermath. Land prices dropped dramatically, and production prices outstripped commodities' market values. Poorly performing economic sectors eroded the base of the national economy, culminating in the Great Depression.

Despite the economic situation, the farm publishing industry remained about the same from 1920 to 1940. Although the numbers of general farm periodicals dropped significantly, total circulation grew from 17 million in 1920 to 22 million in 1940.[22] Perhaps the difficult economic situation made information more valuable to farmers. Subscription rates continued to be inexpensive, and magazine publishers made a strong effort to sell subscriptions. Magazines began to be printed less frequently, and some publishers of periodicals experimented with free, controlled circulation to attract advertisers.

Radio Reaches Out

New communications technologies were introduced to rural audiences during 1920–1940 as well. Radio, movies and popular magazines targeted rural audiences. Although some competition existed, farm periodicals generally welcomed radio. Radio had a profound effect on the lives of rural Americans. Similar to the expectations of some for the World Wide Web, radio's promise was so remarkable that rural people believed it would keep young people on the farm, and diminish the allure of the city. It brought the sounds of the world into their homes. Farm homes in the early 1920s rarely had electricity, but radios were operated on batteries with an antenna wire running from the house to the barn or windmill. Rural Americans gained more than anyone else from radio because of their isolation. Not only were they entertained by the new device, but they also received valuable and timely information about weather and markets.

Prior to radio, farmers relied on newspapers and telephones for information on the weather; these reports were slow and vague. In the early 1900s, the USDA was interested in broadcasting for farmers. The technical bugs were worked out of radio broadcasting as early as 1910, but airwaves still were not available because of an attempted monopoly by the U.S. Navy. However, in the 1920s the government began to realize the potential of radio for nondefense purposes. Herbert Hoover, then secretary of commerce, led a 1922 radio conference in Washington, D.C., that stated that no use of radio, except for military purposes, should supersede use for agriculture. In 1921 WHA in Madison, Wisc., first broadcast weather reports. The 1923 annual report from Secretary of Agriculture Henry Wallace noted that 117 general broadcasting stations and 27 naval stations were broadcasting daily weather reports. In the 1930s and 1940s, more sophisticated weather reports were broadcast in some areas for frost alerts. California citrus growers, Wisconsin cranberry growers and others benefited from these reports.

As they still are today, market reports were critical to farmers in the 1920s. Before radio, farmers generally had to take the word of grain and livestock buyers regarding market prices. Radio changed that. Congress established the Federal Bureau of Markets in 1913, which used telegraph lines to send market quotes to newspapers and farm journals. As radio broadcast technology became available, the Bureau, which was housed in the Department of Agriculture, began broadcasting to radios in a 200-mile radius around Washington, D.C. The first station to broadcast market reports was WLB, part of the University of Minnesota, in February 1921. KDKA in Pittsburgh, Pa., and KFKA in Greeley, Colo., quickly followed suit. By early 1922, 35 stations were licensed to air market information. By 1925 more than 500,000 farmers were able to receive market infor-

Early radio broadcasts greatly benefited farmers. The market and weather reports were faster and more reliable than ever before.

mation, and by 1926, 500 stations were reaching 1 million farm families.[23]

Also during the early 1920s, state agricultural colleges began producing informational programs targeting rural people. In 1922, WOI in Ames, Iowa, broadcast Extension lectures on farm topics. In 1924, William Jardine acquired funding for KSAC in Manhattan, Kan., which began airing regularly scheduled lectures that could be taken as a correspondence course. About 2,000 students from 30 states, Canada and Mexico registered for the course. Jardine left the presidency of Kansas State University and became the U.S. Secretary of Agriculture, bringing with him an interest in broadcasting. Under Jardine, the department started its radio service in 1925, which was led by Sam Pickard. The service initiated educational programs including the "Farm School of the Air" and "Farm Flashes." Other colleges that established programming in the early 1920s include the Massachusetts Agricultural College Extension Service and Iowa State, Michigan State and Texas A&M universities.

Radio pioneers quickly learned that advertising was the key to profitability; however, several commercial stations did not learn that lesson early enough and consequently went out of business. They also learned about giving the audience what it wanted. Farm directors such as Roy Battles (WLW), Art Page (WLS) and Larry Haeg (WCCO) improved the professionalism associated with farm broadcasting, using professional announcers, reporting accurate, timely research and information and avoiding "hayseed" humor.[24]

Many programs for rural audiences on commercial stations at the time combined entertainment with weather and crop reports, soil conservation information, practical farming tips and home economics. One of the most successful was the "National Farm and Home Hour" broadcast from Chicago on WMAQ. It was sponsored by the National Broadcasting Company and the USDA, and featured music and colorful celebrity guests. The extremely popular entertainment segment "The Fibber McGee and Molly Show" originated on the program. Host Everett Mitchell's opening line "It's a beautiful day in Chicago" greeted millions of listeners and became a standard. The program also provided the government with an avenue to reach farmers and present agriculture to the public.

In addition to entertainment programming, religious programs blossomed on radio, and politicians quickly found that radio offered an important way to reach the public; however, it is not clear whether rural voters were better informed on issues. Radio provided an avenue for new types of music, cultural programming and other entertainment. Regardless of other purposes, though, farmers wanted more educational features and market information.

How is agricultural communications different from mass communications? Or trade journalism? Should these programs be housed in schools of journalism and mass communication or in agricultural departments?

Nexus Point 2

Radio became an excellent way for the USDA to discuss its farm programs. In 1933, as Secretary of Agriculture Henry A. Wallace was introducing government farm programs, radio was used to describe the programs intended to support prices, conserve soil, help farmers purchase land and extend electrical lines. Through surveys, the USDA determined that most farmers who signed up for the programs heard about them through radio. Radio, especially the "National Farm and Home Hour" program, also was used to introduce the new agricultural agencies, including the Agricultural Adjustment Administration, the Soil Conservation Service, the Farm Security

Administration, and the Rural Electrification Administration. In 1931, NBC and the USDA expanded programming with the "Western Farm and Home Hour," which was broadcast until the "National Farm and Home Hour" expanded westward in 1938. After a lengthy battle between NBC and the USDA regarding agricultural content, the "National Farm and Home Hour" ended in 1944, after a 16-year, 4,700-broadcast run. It had hosted nearly every agricultural leader of the time. The program was resurrected in 1945 with Allis Chalmers as a sponsor with the USDA. Milton Eisenhower, a former USDA director of information, was not supportive of the USDA's participation with private company sponsorship, because he was concerned about the bias this implied.[25]

Broadcasting expanded once again in the 1940s and 1950s with the introduction of television. The USDA again took advantage of the new medium. In 1954 Layne Beaty became chief of the USDA's Radio and Television Service, a post he held for more than 25 years. In 1961 he began the USDA's work in television with what would become known as "Across the Fence." In 1962 the USDA started "Down to Earth," a 4.5-minute agriculture spot to be incorporated into news programming. Both programs lasted into the late 1970s. Although television became immensely popular with the general public, farm programs had moderate success. In some markets, farm reports were incorporated into noon news shows, whereas in others a farm show stood on its own. The "U.S. Farm Report" was (and is) one of the most successful.

Seeds of Change

Despite the introduction of radio, farm magazine publishing continued to thrive after 1940. From 1940 to 1955, the total number of periodicals in the United States declined,but farm magazines increased from 300 in 1940 to 390 in 1955, and circulation numbers increased even more. About 80 percent of these publications were still general in content. Specialized farm publications grew in number, but only slightly faster than general farm publications. Farmers were a strong buying market during this period. Farm output grew despite fewer farms and little increase in cropland. Output was in demand to support World War II supplies, postwar relief and the Korean Conflict. With better educations and in need of information to support production demands, farmers became more reliant on farm publications for information on innovations.

Until 1950, farm publications attracted advertising from farming sector industries as well as advertising for general consumer goods. However, consumer advertising began to wane. Agricultural advertisers wanting to reach the most qualified buyers (farmers) began to focus their dollars on more specialized publications and audiences. The amount spent to reach each farmer increased although total farm advertising in periodicals decreased. Advertisers were using a more diverse portfolio of communication methods in their advertising mix to reach farmers, relying less on farm periodicals and more on direct mail advertising.

House organs, or company-produced magazines, saw tremendous growth during this period, creating greater demands for the public relations component of agricultural communications. Agricultural house organs increased 50 percent during the late 1950s. These publications allowed a company to tell its own story, dominate the medium, avoid competition within a commercial publication and reach only potential buyers. House organs maintained editorial content, which added to their credibility, and provided a way to publicize company research, which was growing. Bulletins from private companies, similar to those produced by agricultural colleges, also appeared.

Agricultural magazine publishers became concerned about their publications being replaced by house organs and about other industry issues. Several large farm magazines suffered circulation losses and some went out of business, creating concern among most agricultural publishers. Publishers responded by increasing subscription rates as circulation declined. They also trimmed costs by reducing their editorial staffs, page sizes and frequency of publication. Publishers also worked to draw in more advertisers and more purchases from current advertisers by creating ways to reach more-specialized farm audiences. Advertisers saw these specialized audiences as representative of more-qualified buyers, a better targeted market.

Advertisers were offered more services by publishers too, such as split print runs targeted to different geographic regions and readership studies. Publishers sold mailing lists to agricultural marketers.

In the general farm publications, subject matter became more specialized. Gone were fiction, cultural pieces and even the sections devoted to home and family. To reach more targeted markets, publications began specializing by subject matter (beef cattle, row crops, swine, etc.), geographic region and reader buying power, with new publications targeting farmers with higher economic status. Free, controlled circulation sped specialization, and from 1950 to 1970 the number of specialized farm publications tripled.[26]

Agricultural media as a whole gained sophistication from 1950 to 1970 and required improved quality in the work of communications specialists at agricultural colleges and universities. In addition, larger budgets at these institutions resulted in the hiring of more agricultural researchers and Extension specialists, who produced more numerous scientific papers, trial results and bulletins, and needed more support from communicators. The prosperity in agriculture was not just on campus.

> ***Readers will increasingly require more specialized reporting and clear, understandable interpretation of complex ideas, trends, concepts and technology. Readers want it all: To have access to lots of choices and opportunities, but also the means to make simple, confident choices. Communicators must increasingly be both authorities in subject content as well as excellent interpreters with a point of view on that content.***
>
> **—*Loren Kruse,* Successful Farming**

Just before 1970, the Pentagon's Advanced Research Projects Agency began linking or networking computers to form ARPANET, which grew rapidly in the 1970s and 1980s. In 1986 the National Science Foundation formed NSFNET, which eventually became the Internet. Meanwhile, in Switzerland, Internet protocols were being developed that would become the basis for the World Wide Web. Although the effects of these new electronic technologies were not immediately used in the mass media, their ramifications certainly are being felt today.

The 1970s also were a period of industry prosperity. When they were required to pay, readers responded to the changes in publishing and advertising by subscribing to more periodicals and paying higher subscription rates. The average farm in 1970 received seven farm periodicals. Commodity prices during this period were strong, and the price of land was high; consequently, land became greatly overvalued.

Because of high inflation and overvalued land prices, farmers increased their debt load significantly. This teetering economic situation came to a crash in the 1980s, with what has become known as the farm crisis. More than the numbers of farmers dropping from the industry, an even greater impact was felt in

> ***Over the decades, many newspapers dropped their agricultural beats as their farm readership withered. . . . Many newspapers are getting interested in agriculture again, but from a different perspective than when they wrote for farm readers. Now many reporters write about agriculture for consumers, who are a much bigger audience. Everybody eats. Many agricultural reporters today spend much of their time writing about food safety, the environment and nutrition.***
>
> **—*Scott Kilman,* The Wall Street Journal**

the publishing industry due to the tremendous decrease in buying power and available income of remaining farmers. In response, advertisers became even more focused on reaching specialized groups and withdrew some advertising dollars from use in their publications. With fewer advertisements, farm publications became smaller, and free, controlled circulation became even more prevalent. In the early 1800s, about 70 percent of the population of the United States worked on farms. By the 1990s, less than 2 percent of the population was farm based. This affected the way that agriculture was reported to the general public as well. Major metropolitan dailies dropped coverage of agriculture decades earlier and by the end of the 20th century were covering it from consumer and environmental viewpoints.

The farming sector was changing as well. With trends continuing through the 1990s, the number of medium-sized farms continued to decline whereas the number of large farms grew, particularly among those that sought the legal and economic advantages of incorporation. Small farms, or hobby farms, also increased. Segmentation in publishing followed the demographic trends, with publications specializing to meet the needs of their target audiences. The late 1990s witnessed commodity price

What is the administrative home of your agricultural communications or agricultural journalism program? List two or three advantages and disadvantages of housing these programs within larger departments with other academic programs as opposed to placing them in their own departments.

Nexus Point 3

declines that challenged the entire industry. As the agricultural industry changes, agricultural communications sits on the cusp of change too.

Notes

1. Paskoff, B.M. (1990). History and characteristics of agricultural libraries and information in the United States. *Library Trends 38*(3), 331–49.
2. Crawford, N.A. and Rogers, C.E. (1926). "Agricultural Journalism." New York: Alfred A. Knopf.
3. Fusonie, A.E. (1988). The history of the National Agricultural Library. *Agricultural History 62*(2), 189–207.
4. Paskoff, "History and Characteristics."
5. Marti, D.B. (1980). Agricultural journalism and the diffusion of knowledge: the first half-century in America. *Agricultural History 54*(1), 28–37.
6. Paskoff, "History and Characteristics." The Berkshire model was emulated by numerous other fairs for its ability to attract those with smaller farms.
7. Demaree, A.L. (1941). "The American Agricultural Press: 1819–1860." Morningside Heights, NY: Columbia University Press. The author also notes that dirt farmers were heavy users of almanacs, which contained astronomical data as well as various sorts of rural information.
8. From address by Abraham Lincoln to the Wisconsin State Agricultural Society at its annual fair on September 30, 1859. Reprinted in Demaree, "The American Agricultural Press," p. 305.
9. Marti, "Agricultural Journalism."
10. Burnett, C. and Tucker, M. (1990). "Writing for Agriculture: A New Approach Using Tested Ideas." Dubuque, Iowa: Kendall/Hunt Publishing.
11. Skinner, J.S. (1819). *American Farmer, I,* 5.

12. Paskoff, "History and Characteristics."

13. Maixner, E. (1999). A spirited start. *Ohio Farmer 295*(3), 42.

14. Marti, "Agricultural Journalism."

15. Schudson, M. (1978). "Discovering the News: A Social History of American Newspapers." New York: Basic Books.

16. Ibid.

17. Evans, J.F. and Salcedo, R.F. (1974). "Communications in Agriculture: The American Farm Press." Ames: Iowa State University Press.

18. Marcus, A.I. (1986). The ivory silo: farmer-agricultural college tensions in the 1870s and 1880s. *Agricultural History 60*(2), 2–36.

19. Paskoff, "History and Characteristics."

20. Duncan, C.H. (1961). "Find a Career in Agriculture." New York: G.P. Putnam and Sons.

21. Evans and Salcedo, "Communications in Agriculture."

22. Ibid.

23. Wik, R.M. (1981). The radio in rural America during the 1920s. *Agricultural History 55*(4), 339–50.

24. Baker, J.C. (1981). "Farm Broadcasting: The First Sixty Years." Ames: Iowa State University Press.

25. Ibid.

26. Evans and Salcedo, "Communications in Agriculture."

2

The New Age of Agriculture

Since the dawn of time, or at least since the time when the first humans observed that if they placed a grain into the soil, added water and waited, a plant would grow, or observed animal husbandry in action, agriculture has been a part of the human experience. Throughout history agriculture, in many forms, has been present. Creationists would argue that agriculture was first observed in the Garden of Eden; evolutionists would note that agriculture came as humans, in their early stages, learned to herd and slaughter animals for food and grow plants for sustenance. Regardless of its beginnings, agriculture has been a dominant cultural influence. It has been a way of life that has had a great impact on humankind today. Anthropologists describe early human cultures as "hunters" and "gatherers." These early societies were the first farmers—gathering or harvesting, hunting or gleaning for food.

Until the 19th century, agriculture in the United States shared the history of European and colonial areas and depended upon Europe for seed, stocks, livestock and machinery. That dependency, especially the difficulty in procuring suitable implements, made American farmers somewhat more innovative. They were aided by the establishment of societies that lobbied for government agencies for agriculture, the voluntary cooperation of farmers through associations and the increasing use of various types of power machinery on the farm. Government policies traditionally encouraged the growth of land settlement.

For years the life of a farmer was thought of as an idyllic one. Living on the land. Serving as a steward of the soil. Providing food for America's tables. As society moved from an Agricultural Age to an Industrial Age, society shifted from rural to urban, and farmers became less an item of interest for the public. The Homestead Act of 1862 and the resettlement plans of the 1930s were the key agricultural legislative acts of the 19th and 20th centuries.

Agriculture, including its crops, livestock and production practices, has changed dramatically due to mechanization and demands of the public. A strong back was once one of the occupation's major requirements.

Strong analytical skills now are greatly in demand.

Agriculture and the Great Depression

One of the greatest periods of uncertainty for U.S. agriculture came shortly after the turn of the century.

Americans had known times of prosperity between the Civil War and World War I, but after World War I, the economy took a turn that sent Americans, especially people in rural communities, into great turmoil. During World War I, federal spending grew three times larger than tax collections, and when the government began to cut spending in an effort to balance the budget, a recession began. What followed during the 1920s included bank failures (600 or more each year); depression in agricultural, energy and coal-mining sectors; and drops in farmland value of 30 percent to 40 percent. A shift from human resources in industry to more automatic and semiautomatic processes created widespread unemployment. All of these events and more seemed to culminate in one historic event: the crash of the stock market on October 24, 1929. Agriculture, which was already in recession, felt the sting. In 1929, for example, the annual U.S. per capita income was $750; for farmers it was $273. Farm prices fell 53 percent between 1929 and 1933. Although there have been other stock market disparities of even greater financial value, this one event helped shape the policy and history of future decades. It left a greatly changed America in its wake.

An ecological and human disaster that took place in the Great Plains in the 1930s, the Dust Bowl, added to the plight of farmers during the Great Depression. The Dust Bowl included parts of Kansas, Nebraska, Oklahoma, Texas, New Mexico and Colorado—some of the most productive cropland in the United States. It was caused by longstanding agricultural practices that left the soil exposed to erosion, particularly by wind during periods of drought.

Beginning in the early 1930s, the region suffered a period of severe drought, and the soil began to blow away. The organic matter, clay and silt in the soil were carried great distances by the winds—

in some cases darkening the sky as far as the Atlantic coast—and sand and heavier materials drifted against houses, fences and barns. Millions of acres of farmland became useless, and thousands of people were forced to either leave their homes or settle further into financial strife.

The Dust Bowl lasted about a decade. Beginning in 1935, intensive efforts were made by both federal and state governments to develop adequate programs for soil conservation and rehabilitation of the Dust Bowl. Measures taken included seeding large areas in grass; rotating wheat, sorghum and lying fallow; introducing terracing and strip planting; and planting long "shelter belts" of trees to break the force of the wind.

Farmers in some of the nation's most productive agricultural areas were hit hard by the Dust Bowl, especially because economic conditions across the country were weak due to the Great Depression.

Franklin Delano Roosevelt was considered the president who brought the United States out of the Depression through revolutionary government programs. His New Deal legislation benefited agriculture and rural America at a time when land values and farm income were incredibly low.

Roosevelt was inaugurated in March 1933, and during his first 100 days in office he took actions that sought immediate relief for Americans.

Although much more information about the general policies and legislation associated with the New Deal can be found in history books, this chapter discusses only the agencies and legislation that specifically affected farmers and rural Americans.

The Agricultural Adjustment Administration was created to address overproduction and low prices. This administration gave subsidies to farmers who curtailed their production. This move to pay farmers not to farm is one that the U.S. Department of Agriculture works with even today. The Civilian Conservation Corps provided employment to young, unmarried men. The Corps planted trees, built flood barriers, fought forest fires and maintained forest roads and trails. Congress also passed the Farm Credit Act of 1933, which refinanced one-fifth of all farm mortgages for a period of 18 months and created the Farm Credit Administration. The Farm Credit Administration was created to regulate, coordinate and examine the institutions belonging to the Farm Credit System, a system of banks and associations that made loans to farmers, ranchers and agricultural cooperatives. The Tennessee Valley Authority is one of the largest and most successful programs to come out of the New Deal. It brought electricity, flood control and improved navigation and living conditions to the Tennessee Valley.

In 1935 the Rural Electrification Administration (REA) was created to raise the standard of rural living and to stop the migration of rural Americans into the cities. After the REA was created, 9 of 10 farms received electricity for the first time. Historians observe that, ironically, rather than slow down migration into the cities, electrification actually sped migration because countless farmhands were replaced by machines.

In 1936 Congress passed the Soil Conservation and Domestic Allotment Act in response to the Supreme Court's decision that part of the Agricultural Adjustment Administration was unconstitutional. The court ruled against the payment of a subsidy to farmers to curtail production. The new act allowed the administration to pay subsidies to farmers who planted soil-enriching rather than staple crops. This redressed the grievances of tenant farmers and sharecroppers who had been hurt under the previous system.

In 1938 Congress passed the Agricultural Adjustment Act, in response to the Supreme Court's 1936 decision that allowed the Agricultural Adjustment Administration to determine acreages for staple export crops and award loans according to stored surplus crops.

Legislation, agencies and programs all were developed during the 1930s to address erosion of soil and depletion of water resources, rural poverty, rural isolation and the vast economic problems rural America was facing as a result of the depressed economy. These events are historically significant for American agriculture because they laid the groundwork for legislation and government involvement that, in some cases, is still in place today. Changing societal and economic needs have created new federal legislation and government agencies that American farmers must contend with today.

Agriculture After the New Deal

A significant outcome of the New Deal to agricultural communicators is the development of radio. When the REA put electricity into rural homes, farm radio came into those homes. By 1922 the U.S. Department of Agriculture (USDA) reported that 36 radio stations had been licensed by the Commerce Department, and 35 had been approved to broadcast USDA market news. But, most rural Americans had to read market news because they had little contact with broadcast radio. Rural electrification meant that more farmers could hear up-to-date market information on a home radio set. The 1930s saw more stations employing full-time farm reporters. USDA and land-grant college Extension personnel were on the air with farm information. Commercial sponsorship was developing (see Chapter 1).

Nexus Point 4

For various reasons some people do not have access to information technologies. What happens to people who cannot get information via electronic media such as the Internet or CD-ROM? As an agricultural communicator, how can you plan an information campaign that includes everyone?

In the 20th century, steam, gasoline, diesel and electric power came into wide use. Chemical fertilizers were manufactured in greatly increased quantities, and soil analysis was widely employed to determine the elements needed by a particular soil to maintain or restore its fertility. Another outcome of New Deal legislation was the attention paid to natural resources. The loss of soil by erosion was extensively combated by the use of cover crops (quick-growing plants with dense root systems to bind soil); contour plowing (the furrows follow the contour of the land and are level rather than running up and down hills and provide channels for runoff water); and strip cropping (sowing strips of dense-rooted plants to serve as water breaks or windbreaks in fields of plants with loose root systems).

The period following the Great Depression also saw great improvements in the science of agriculture. Selective breeding produced improved strains of both farm animals and crop plants. Hybrids (offspring of unrelated varieties or species) of desirable characteristics were developed; especially important for food production was the hybridization of corn in the 1930s. New uses for farm products, by-products and agricultural wastes were discovered. Standards of quality, size and packing were established for various fruits and vegetables to aid in wholesale marketing. Among the first to be standardized were apples, citrus fruits, celery, berries and tomatoes. Improvements in storage, processing and transportation also increased the widespread marketability of farm products. The use of cold storage warehouses and refrigerated railroad cars was supplemented by the introduction of refrigerated motor trucks, rapid delivery by airplane and quick-freeze preservation, in which produce is frozen and packaged the same day it is picked. Today freeze-drying and irradiation also have reached practical application for many perishable foods.

The history of agriculture in the United States since the Great Depression has been one of consolidation and increasing efficiency. From a high of 6.8 million farms in 1935, the total number declined to slightly less than 2 million by 1999. The average farm size in 1935 was about 155 acres; in 1997 it was 471 acres. In 1997 the area devoted to farms occupied 968 million acres.

According to the 1990 U.S. Census, about 4.6 million people lived on farms. This number is based on a farm definition introduced in 1977 to distinguish between rural residents and people who earned $1,000 or more from annual agricultural product sales. The farm population continues to be a declining share of the nation's total; in 1990, about 1 person in every 54, or 1.8 percent, of the nation's 250 million people were farm residents.

Farm Policy—A Shifting Tide

During the administrations of Presidents John F. Kennedy and Lyndon B. Johnson in the 1960s, the Department of Agriculture made control of overproduction a primary goal of farm policy. Farmers were offered payments in what amounted to a rental for part of their land that would be taken out of production the following year. At the same time, measures were implemented to expand export markets for agricultural products. During this period, the ration of a farmer's per capita income to that of a nonfarm person increased from about 50 percent to about 75 percent.

Direct subsidies for withholding agricultural land from production were phased out in 1973, as a result of a proposal by President Richard M. Nixon. Net farm income swelled to $33.3 billion that year.

Poor grain harvests throughout the world, particularly in the USSR, prompted massive sales of U.S. government–owned grain reserves in the 1970s. Global climatic conditions also helped keep worldwide demand for U.S. produce high through the mid-1970s. Soon, however, exports decreased, prices dropped and farm income began to fall, again without a corresponding decrease in costs of production. In 1976, U.S. net farm income fell to $18.7 billion.

In 1978 a limited, voluntary output restriction was begun by President Jimmy Carter. Called the Farmer-Held Grain Reserve Program, the action took grains off the market for up to three years or until market prices reached predetermined levels. The program also was intended to provide adequate reserves of essential commodities such as corn and wheat, lessen food-price gyrations and combat inflation, give livestock producers protection from extremes in feed costs and contribute to greater continuity in foreign food aid.

On January 4, 1980, President Carter declared a limited suspension of grain sales to the USSR in response to its invasion of Afghanistan. Despite the grain embargo, the United States continued to honor a five-year agreement already in effect that committed it to sell 8 million tons of grain to the USSR annually. Despite efforts by President Carter's opposition to void the embargo in 1980, an election year, it remained in effect. Administration officials argued that the Soviets had never been a major customer or even a reliable buyer. American farmers maintained, however, that the action was taken at their expense and made 1980 one of their worst years. In fact, U.S. farm exports in 1980 reached an all-time high of $40 billion, but the continued rise in costs of production and an extremely hot summer with accompanying droughts affected many farmers adversely. A new crop insurance program, passed by Congress in the fall of 1980, offered relief from such conditions rather than forcing farmers to rely on disaster loans. That year disaster loans amounted to $30 million for feed alone.

When President Ronald Reagan took office in 1981, he lifted the embargo and extended the agreement that allowed the USSR to purchase 8 million tons of grain yearly from the United States. In 1983 the two nations signed a new five-year agreement that obligated the USSR to import a minimum of 9 million tons of U.S. grain annually. Since the breakup of the former Soviet Union, the United States has continued to sell large amounts of wheat to Russia as well as China.

In 1980 a report based on projections by the U.S. government stated that in the next 20 years, world food requirements would increase tremendously, with developed countries requiring most of the increase, and that food prices would double. Less than five years later, however, the U.S. farmer was enveloped in a major crisis caused by exceptionally heavy farm debts, mounting farm subsidy costs and rising surpluses. Many farmers were forced into foreclosure. In the early 1980s, land prices became greatly overvalued. Farmers were able to borrow greater amounts of money based on using their land as collateral. When land values corrected, farmers were left holding heavy debt loads.

The ailing Farm Credit System, a group of 37 farmer-owned banks under the Farm Credit Administration, appealed to the government for a $5 billion to $6 billion fund that would keep the system solvent despite the weak national farm economy. After initial resistance, President Reagan signed legislation in December 1985 creating the Farm Credit System Capital Corporation, to take over bad loans from the system's banks and assume responsibility for foreclosing or restructuring distressed loans.

President Reagan also signed the Food Security Act of 1985, legislation designed to govern the nation's farm policies for the next five years, trim farm subsidies and stimulate farm exports. In the early 1990s, although farms generally still struggled to be economically viable, fewer farms faced the kinds of crises prevalent in the 1980s. Government assistance to farms steadily decreased. For example, from 1987 to 1992, the number of acres placed in federal commodity programs, which pay farmers to leave acres uncultivated, decreased by nearly 84 percent. Furthermore, fewer farms were threatened with bankruptcy. Record-setting production of corn, soybeans, cotton and rice took place in 1995.

In 1996 the U.S. Congress passed the Federal Agricultural Improvement and Reform Act (FAIR), which was signed into law by President William J. Clinton. The act, which will remain in effect until 2003, was the most far-reaching federal farm program in 60 years. FAIR, also called the Freedom to Farm Act, replaced existing farm price-support programs with a system of transition payments designed to let both domestic and export markets, rather than the government, determine crop production. It also increased farmers' flexibility to plant crops to meet world market demands.

In addition, FAIR included food and nutrition, rural economic development and agricultural market-promotion programs; phased out dairy subsidies; and provided incentives to encourage farmers to preserve wetlands and conserve noncrop-producing acreage.

In 1998 and 1999 Americans were faced once again with an agricultural crisis. Climatic disasters, from excessive rainfall to scorching drought, created problems with planting and growing seasons in some parts of the country. Other farmers produced record amounts of grain, creating an oversupply that caused commodity prices to drop to all-time lows. Pork production levels, high due to excellent grain and feed prospects, were so high that the pork market bottomed out. Congress passed a $6 billion farm disaster and market loss assistance bill that included special assistance payments and a tax relief package to serve as a safety net for farmers.

Science and Agriculture

Science has played a critical role in agriculture. Scientific methods are applied now to pest control, limiting overuse of insecticides and fungicides and employing more varied and targeted application techniques. New understanding of significant biological control measures and the emphasis on integrated pest management make possible more effective control of certain types of insects.

Chemicals for weed control are important for a number of crops. The increasing use of chemicals for the control of insects, diseases and weeds, however, has resulted in additional environmental problems and regulations that place strong demands on farmers' skills. These same issues also have created concerns regarding the interface between rural and urban societies. Urban communities are claiming land once used for agricultural interests, and current agricultural operations are finding that they are being restricted because of urban interests and influx. One prime example of this rural/urban conflict is the positioning of animal confinement facilities. As cities expand, such facilities must account for waste and odor issues that were not controversial when the facilities were first planned and built.

Since the 1970s, high-technology farming—including new hybrids for wheat, rice and other grains; better methods of soil conservation and irrigation; and the growing use of improved fertilizers—has led to the production of more food per capita. This trend is occurring not only in the United States but in much of the rest of the world as well. The United States is the world's principal exporter of agricultural products. In 1997 the value of agricultural products exported was about $67.4 billion.

U.S. farmers have the advantage of superior private and government research facilities to produce and perfect new technologies. The USDA has two agencies primarily focused on agricultural research: the Agricultural Research Service (ARS) and the Cooperative State Research, Education, and Extension Service (CSREES).

The ARS is the principal in-house research agency of the USDA. Its research facilities are located near Washington, D.C., as well as throughout the United States, and its researchers work on all aspects of agricultural science. CSREES works in partnership with researchers located throughout the United States in the land-grant university system. Its National Research Initiative provides millions of research dollars each year for study in the biological, physical and social sciences on issues related to agriculture and natural resources. Another component of CSREES, the Cooperative Extension Service, uses nonformal education to provide research information to the public. Throughout the land-grant university system, agricultural communicators are found engaged in the arenas of mass media, electronic media, publishing, graphic design and information technology to extend the research of the federal government and the university to the public.

Private-sector research is also important in today's agricultural climate. Large companies produce the seed stock, germ plasm and genetic refinements offered for public consumption in high-technology laboratories and often in cooperation with major research universities. Agricultural communicators also work to promote research findings to potential users.

New applications of technologies in the 1990s are improving crop production further. Precision farming, also known as prescription farming, site-specific farming or variable rate farming, utilizes global positioning systems (GPS) and geographic information systems (GIS) in the satellite collection and transmission of data to create yield maps of fields during harvest. Farmers use the yield maps as they plant and fertilize their crops the following season. This increases crop production while reducing the use of both fertilizers and fuel. GPS also help farmers comply with environmental regulations when applying fertilizers and pesticides.

Biotechnology also is increasing agricultural productivity. New hybrid corn seed recently devel-

> ***The agricultural communications industry will change in the future because the agricultural industry itself is changing dramatically. Technology like biotech, precision farming and others are dramatically changing agriculture and making it more complex. There is more "information" but perhaps less "knowledge" than ever before in agriculture and that is a huge opportunity for professional communicators. At the same time, the emergence of the Internet as a communications medium, as well as the convergence of many other traditional communications vehicles like print, radio, television and database marketing, has a direct impact on how we as agricultural communicators approach our jobs.***
>
> ***—Jeff Altheide, Gibbs & Soell Communications***

oped to resist the corn borer, an insect poisonous to corn, and improved varieties of barley with disease-resistant genes are now under cultivation. In 1993 the Food and Drug Administration approved a biotechnological product for use in animals, bovine somatotropin (bst). This drug, a growth hormone, is injected into dairy cows. It supplements the natural bst a cow already produces and increases milk production. The use of this drug created concerns among the public about human safety and economy (the oversupply of milk from cows on the hormone). What is not yet known are the potential implications of the long-term use on cows of biotechnological agents such as bst, the effects on humans consuming the milk produced by cows on such agents, or the long-term ramifications to farm structure in the dairy industry.

Cloning, the creation of a living organism identical to another living organism by growing the organism from a single cell from the original organism, found its first applications in agriculture. In 1997 geneticists in Scotland announced that they had, for the first time, cloned an adult mammal (a ewe). Using a technique called nuclear transfer, they removed the nucleus from an egg cell and, with an electric pulse, fused the denucleated egg cell with a whole cell. The electric pulse also stimulates the egg so that it divides, becoming an embryo. The embryo is implanted in a surrogate mother. Dolly, the Scottish ewe, became the focal point for discussion

Nexus Point 5

Agricultural issues are among the most important facing consumers today—food safety, environment, water quality, urban sprawl. But some have raised questions about the willingness of agricultural editors and news directors to address controversial issues in their media. How are controversial issues covered in the agricultural media? Should agricultural news and information be reported more like general news?

worldwide regarding science and ethics. Implications for cloning in crop and livestock production in the United States are quite important. In 1999 cloning was considered an option for providing animals for food in Third World countries.

Changing Face of Agriculture

Today, agriculture holds a unique position in society. As more and more of the world's population moves from rural to urban areas, agriculture as a way of life is changing. Farming no longer dominates rural economies. Today's rural communities are more likely to be dominated by manufacturing, services or government jobs, although farming income still strongly fuels rural economies.

Whereas farming provided more than 14 percent of all rural jobs in 1969, that proportion dropped to about 7 percent by 1993. Recent job growth has occurred in the service sectors of the rural economy. Retirement and recreation activities are particular economic bright spots in the rural landscape. Many rural counties serve as residential areas for workers who commute to jobs in other counties. Only about 20 percent to 23 percent of rural jobs are associated with the food and fiber sector. With the exception of farming, food and fiber jobs are more likely to be located in U.S. metropolitan areas—in wholesale and retail trade.

Agriculture's role in food production has not changed, however, and more and more of the world's food production is centered on agriculture—with fewer and fewer people engaged in production farming. An increasing share of U.S. food and fiber is being produced on fewer farms, and farms have become more specialized.

Experts predict that the number of farms will continue to decline. As we previously noted, there are fewer farms today than 50 years ago. Although the number of acres in U.S. farming has remained about 1 billion over the same time period, the structure of farms has changed. Larger farms (more than 1,000 acres) are increasing in numbers as are small farms (less than 75 acres). Middle-sized farms are declining in numbers.[1]

Because there are fewer farms but larger farms, there is a shift away from family farming. Family farming has long been viewed as an idyllic lifestyle. But, changes in the farm economy during the 1980s forced many farm families to look closely at the lifestyle they were leading. This period saw many farm couples unable to meet their financial demands, and, consequently, either the husband or wife spent time away from the farm in order to earn additional income. However, the shift toward earning a nonfarm income never really returned to the family farm. Children raised on family farms shifted their interest in agriculture away from production agriculture to agriculture through agribusiness or other career opportunities. They now find employment in crop and livestock consulting, sales, marketing and wholesale and retail agriculture.

An increasing share of U.S. food and fiber is being produced on fewer and fewer farms. According to Census of Agriculture figures in 1992, 3.2 percent of all farms were producing one-half of all sales of agricultural commodities from U.S. farms. That means that larger commercial farms are primarily responsible for the production of most of the food in the United States. The trend toward increasing the concentration of production among fewer farms is expected to continue into the 21st century. Large commercial farms are expected to increase in number and will likely be 3.5 percent of all farms by the end of the 1990s. The number of small commercial farms is expected to decline.

Agriculture is big business. Agricultural business is dominated by large multinational organizations with business interests reaching around the world. These companies, such as Farmland Industries, Archer Daniel Midland and ConAgra, have worldwide operations, vertically integrated from farm to retail outlet. Many of these companies contract with small farmers to produce one component of the product needed to develop an agricultural enterprise. Others own farms and manage the production of the product they will ultimately brand and sell.

> ***Even though the farm population is dropping, agriculture is still the wellspring of roughly one-sixth of the economy. . . . Among other things, the agricultural sector is a big part of the U.S. export picture. It supports much of the banking industry and drives the fortunes of many manufacturers. Some of the world's biggest and most powerful companies are tied to agriculture.***
>
> **—*Scott Kilman,* The Wall Street Journal**

Even with all of these changes—fewer farmers, more very small farms, large farms producing more of the agricultural products, rural America moving away from farming—America still has a large interest in agriculture and farming.

Today's average farmer is extremely hard to describe. Many rely on nonfarm income to supplement their farm-generated income. As consumers, farmers of course are interested in agricultural products and services, but they also are interested in products not generally attributed to agriculture. As consumers, they have become more similar to other groups in American society. The gap between rural and urban society has narrowed. Media and popular culture have changed the way that farmers work, live and interact with other parts of society.

Agriculture and Technology

Drive on to a farmer's property today, and you will quickly notice the innovation for which agriculture is known. Farmers are like any other people engaged in business and, as such, have the resources at hand to succeed in business. Depending on the commodity in which a farmer specializes, technology is critical for efficiency and fiscal success. Some farming operations require acres and acres of crops planted from fencerow to fencerow. Corn, wheat, soybeans and sorghum require equipment, planters, harvesters, tractors, irrigation systems and other production tools featuring the highest technological advances. Computer technology is built into most equipment in use on farms today. Specialty crops and specialty farming operations are also tied into advanced technology.

Livestock production with artificial insemination techniques in order to produce high-quality meat and dairy production with electronic milking systems are prevalent. New technologies allow producers to track a steer from gate to plate, accounting for the feed it consumes, the antibiotics it is given and the quality of its carcass. And, as noted previously, precision farming, or "site-specific farming," that tailors soil and crop management to match conditions at every location in a field is also part of the agricultural picture. Designer crops and livestock tracking will allow consumers to buy products that are identity marked. If he or she wants, the consumer can trace purchased meat back to the animal from which it came.

Technology is not limited entirely to the production aspects of agriculture. Farmers, as has much of society, have adopted personal computer technologies into their farm operations. Agribusiness managers predicted that in 1998, 80 percent of commercial producers would be using a computer for farm management.[2] According to USDA data from 1997, 44 percent of producers with sales receipts of $10,000 or more had computers. This percentage increased to 60 percent for producers with sales receipts of $100,000 or more.[3] Computers track feed rations and market trends and provide access to other rural industries for farm interests. Farmers also have adopted satellite technologies, allowing them access to worldwide media, information and entertainment services.

The agricultural industry and allied businesses are broader than production agriculture, and the industry is being held accountable by the public for the quality and safety of its products from "gate to plate."

Agriculture's New Age

Today, agriculture is being held accountable for environmental quality and food safety because these are things that are important to society. Consumerism is a critical factor affecting agriculture. Recent media reports about food safety have brought agricultural production into question. Mad Cow disease, Alar on apples, beef tainted with *Escherichia coli,* botulism—all are highly visible issues focused on food production by America's farmers.

> ***Agricultural promotions is changing rapidly. The change is due to the changing nature of the food and agricultural industry. And potential customers are more "worldly" in their perceptions. . . . Customers are more skeptical of advertising claims and offers. They know their needs and expect more than pretty pictures.***
>
> ***—Donna French Dunn,***
> ***American Agricultural Economics Association***

Agriculture is an industry that ebbs and flows with changing weather patterns, global economies and national trends and fads, and it suffers from a lack of understanding. Politically, agriculture has lost some footing with a shift from rural to urban population centers. Advocates for agriculture must come from the food industry sector, not primarily the agriculture sector.

Even the names of colleges of agriculture have changed to be more inclusive of agricultural functions that are not specifically related to production agriculture. Today it is not uncommon to see a College of Food and Agricultural Sciences or an Institute of Food Safety and Natural Resource Development.

> ***Agriculture and society will need new and improved agricultural information channels and services that are geared to the scientific, progressive, change-oriented dimensions of a culture. At the same time, agriculture and society will need a communications system that recognizes and maintains the stabilizing, deep-rooted human and social dimensions of a culture. The frictions will be tremendous because we are dealing with human values in conflict.***
>
> ***—Jim Evans, professor emeritus, University of Illinois***

Unlike previous generations, most Americans do not understand how food is produced, which results in a lack of confidence in the safety and quality of the food supply. A typical response for most people is to distrust that which they do not control. Because consumers do not control production agriculture, they tend to unduly fear it. Agriculturists face an educational challenge.

Notes

1. The USDA defines a *farm* as one that produces $1,000 or more per year from agricultural products.

2. This prediction was made by agribusiness managers in a Purdue University study that was reported in the October 1998 issue of *Agrimarketing Magazine.*

3. Farm computer usage and ownership, released July 1997 by the National Agricultural Statistics Service, USDA.

3

The Age of Choice

Americans today live in a world of choice. If someone from the former Soviet Union or a Third World country visited a supermarket in the United States, he or she would be overwhelmed by the choices to be made. Just as there are choices in food or fashion, the choices Americans, in general, and farmers, specifically, can and do make in their search for information, entertainment and business are important to future agricultural communicators and journalists. Consider the following scenario:

> John Doe is a grain and livestock farmer in central Illinois. He farms 600 acres of corn, soybeans and wheat; annually raises 50 to 60 head of Angus cattle; feeds out more than 100 hogs; and annually rotates between sunflowers and safflowers as alternative crops. John is married and has two children. His wife, Jane, teaches English at the local high school. Their children, twins aged 18, plan to attend the local university in the fall, majoring in business administration and graphic design. Neither plans to return to farming with John.
>
> Each morning John arises at 5:30 a.m. He grabs a bite of breakfast while watching "AgDay" on WCIA, CNBC for a summary of the markets from the previous day, and then "CNN Headline News" for the world and national news highlights. Jane follows

him into the kitchen and, as he leaves the house, changes channels to NBC's "Today Show" and the local NBC affiliate for local news. She sits at the computer terminal, logs onto America Online (AOL) and reads e-mail from family and friends.

John returns to the house after a morning of working in the farm shop and discing fallow ground for planting. He has spent the first part of the morning listening to WILL, the state's public radio affiliate, enjoying National Public Radio's "Morning Edition." After that program concluded at midmorning, he turned to KLMN, his local radio station. John enjoys their mix of news, music and talk radio. Upon entering the house, John boots up his DTN terminal and checks the markets in Chicago, Kansas City and worldwide. He also looks at news from the Illinois Cooperative Extension Service; he is concerned about a bean leaf beetle outbreak he heard about over coffee in town that morning. John logs onto his AOL account to check on e-mail from a college friend; while on-line he moves to "Farm Journal Today," one of his bookmarked Web sites, where he finds news on the bean leaf beetle problem as well.

John returns to the field for an afternoon of discing and more country music and news. When he returns to the house, he kicks back in his easy chair and, with the local Fox affiliate on the television, skims through his new copies of *Farm Journal* and *Illinois Farmer* magazines. His son sits at the family computer surfing the Web for college sites, and his daughter prepares dinner from a recipe she took from the Healthy Cooking CD-ROM the family received as a holiday gift.

This fictitious scenario reflects a major theme of this chapter: the notion of a "media mix," the combination of media sources to which a consumer turns to for information. If you remove the identity of John and Jane Doe and their children as a family living on a farm, earning a farm and nonfarm income, you might imagine that this media mix might belong to any family in the United States. In fact, that is the point of this scenario: to illustrate that what was once considered an audience that turned to specialty media alone—agricultural media—cannot be considered as such today.

John Doe and his family have information needs as well as entertainment needs; they are meeting some of those needs via a unique mix of media sources—some tied specifically to agriculture and some considered mainstream or with little tie to agriculture. As observed in Chapter 2, fewer farmers are involved in agriculture today; those who are often look outside agriculture for their information needs.

Farmers, however, are an important target market, each year purchasing farm equipment, feed, seed, pesticides and other farm chemicals, consultant services and other agricultural products. So, in being a target market, agribusiness, in particular, is interested in reaching them.

> ***The Information Age seems to be bringing with it a general confusion about "information" and "communication." The former is being taken, by some, as a sufficient substitute for the latter. This interpretation offers a tempting, but fallacious, shortcut to human communication, a convenient way to define the communications task as providing access to databases and information sources. It also leads to a dangerous narrowing of perspective about the challenges at hand.***
>
> ***—Jim Evans, professor emeritus, University of Illinois***

Agricultural Media

Writing a chapter on agricultural media 10 years ago would have been much easier than it is today.

Agricultural media, newspapers, magazines, radio and television all represented clear and well-defined entities. Today, because technologies and economics play a greater role in agricultural media, the picture is not quite as easy to describe. The following sections describe broad categories of agricultural media, but they cannot clearly delineate one medium from another. Some media are merged in electronic formats; some have become more "mainstream."

Print Media

Farm publications continue to be the main source of information for farmers. Younger farmers tend to obtain information from publications more often than older farmers, and they have numerous choices of where to get their information. Two of the largest farm publications, *Successful Farming* and *Farm Journal,* blanket the country. These publications are offered via subscription to farmers, and they often target specific producers with specific editions of their publications. For example, a hog producer in Indiana might receive a slightly different publication than a beef producer in Nebraska. Advertising and content would shift slightly depending on the commodity breakdown of the subscriber. The benefits of this method of distribution are better targeting of the reader for information he or she desires and providing better advertising opportunities for companies.

Both *Successful Farming* and *Farm Journal* have moved to electronic versions of their publications, which can be found on the Internet. They also have diversified their publishing into other more specific publications, such as *Farm Journal's Top Producer* and *Pro Farmer.*

One of the longest running controversies in agricultural publishing is the debate over controlled vs. paid subscriptions to farm publications. Controlled or free subscriptions are initiated when a

Who is your audience, specifically? Communication has become less oriented to mass appeal and more focused on audience segmentation.

farmer returns a subscription card or signs up at a farm show and begins receiving the publication. The argument can be made that publishers of controlled subscription publications have little true knowledge of the readership of their publications, thus making

> ***To be trustworthy is becoming the most precious asset of a person or business or any entity. As individuals and businesses are flooded with more information, we increasingly must rely on others to filter and process much of this information for us. Thus, agricultural publications could play an increasingly important role. However, our industry has failed in general to provide enough value in our publications that readers will pay for them. By letting advertisers pay the total bill and readers none, we have forfeited much of our credibility and trust. As trust is lost and agricultural publications lose readership time, these publications could eventually be of less value to agrimarketers too.***
>
> **—*Loren Kruse,* Successful Farming**

> ***Ethics matters—in any field. There can be no "profession" without professional standards, and there can be no meaningful professional standards that ignore ethics. At the end of the day, you want to be able to look at yourself in the mirror and be confident that, in advancing your own cause, you haven't done harm to someone else. In simplest terms, that's what ethics is all about.***
>
> *—Robert Hays, professor emeritus, University of Illinois*

it harder for advertisers and journalists to know whom they are reaching. With a paid circulation, advertisers and journalists can feel that the reader, first of all, values the publication because he or she pays for it, and, second, has a specific interest in the subject matter of the articles or advertisements in the publication. While free, controlled subscription ensures that magazines are targeting specific audiences, thus pleasing advertisers, it also may make the magazines more vulnerable to pressures by advertisers regarding editorial content because these magazines are not getting feedback from readers in the form of paid subscriptions; rather, the most common form of feedback comes from advertisers.

Most states have a "state" agricultural magazine: *Nebraska Farmer, Wallace's Farmer* (Iowa), *Minnesota Farmer, Texas Farmer-Stockman,* etc. Any discussion of these publications must include a trend in agricultural publishing reflecting the structure and ownership of media. At their "birth," most state farm publications were owned locally by small publishing companies. As economies and publishing costs changed, these publications became attractive pieces of property for larger media organizations. One such organization was Farm Progress Publications. Over time, it acquired 17 state farm publications and operated them much like a large corporate entity. This type of media ownership was not limited to just agricultural media— it was happening in media throughout the country.

In the 1980s, Farm Progress Publications, already a subsidiary of Harcourt Brace Jovanovich, Publishers, began moving through a variety of financial acquisitions, being bought and sold. Ultimately, the *Nebraska Farmer,* a state farm magazine (and its 16 companion state magazines) was owned by the Disney Corporation! There were numerous steps and companies in between, and the gyrations of this type of sale created great concern about the magazine's reliability, management and attention to the farm community.

The question also was raised regarding the objectivity of media considering this type of ownership. Would media owned by a corporate giant such as Disney be able to conduct investigative reporting on its parent company objectively? Another member of this acquisition was ABC Capital Cities, owners of the ABC Television Network. Today, the Farm Progress publications are owned by Rural Press Limited.

Whereas national and state publications have had their share of "stresses" in recent history, another group of publications succeeded in meeting its readers' needs. Livestock publications, published by major livestock breed associations and other commercial entities, have a long history of quality publishing. The Livestock Publications Council, a group organized around membership from each of its 125 publication members, provides leadership on current trends and issues in publishing. Member publications focus on specific breeds or livestock classes. Most issues include feature articles, editorials and advertising related to the specific breed or livestock class. The publications appear monthly, quarterly or semimonthly and often serve as the official publication for a particular breed association.

The agricultural trade press, including magazines and newsletters—for example, for feed millers, veterinarians and other specialty careers and businesses—are also an important part of the agricultural communications mix.

Regardless of the type of publication, agricultural businesses continue to see print advertising as an excellent way of reaching producers. In 1998 agribusinesses spent approximately $147 million in print advertising.

Nexus Point 6

What is service journalism? In what ways is it an appropriate description of agricultural journalism? In what contexts is it inappropriate? Do you feel that agricultural journalism is advocacy journalism? Why or why not? If not, should it serve such a role?

Broadcast Media

Electronic media, radio and television, have had mixed success in the United States. Farm radio continues to be of strong value to agriculturalists today. In a study commissioned by the National Association of Farm Broadcasters, farmers indicated that farm commodity market reports are the most important reason they listen to farm radio. Farmers also looked to farm radio for farm news, weather and general information/entertainment. The media mix discussed previously also must factor in the time of day consumers use a particular medium. Radio is the main source of information for farmers in the morning. During the noon hour, television and radio equally share farmers' attention.

Radio, however, is not where agribusiness spends advertising dollars. In 1998, of more than $860 million spent on all advertising, American agribusiness spent only $60 million on radio advertising.

Whereas many radio stations have farm directors, agricultural news and market coverage often is complemented by services of agricultural radio networks, such as the Brownfield Network and Mid-America Ag Network. These networks vary in size and number of station members but provide a focused source of agricultural news and market information.

Agricultural television, in recent years, has had mixed success. Channel Earth Communications was an all-agriculture channel offered to farmers who subscribed to a small-dish satellite network, DIRECTV. Channel Earth began broadcasting in March 1997, offering 13 hours of programming daily. News, weather, local news and features were all part of the coverage mix offered by this venture. Programming during the 13 hours included material provided by the local land-grant universities and colleges.

Placing an agricultural channel on a direct satellite broadcast network seemed like a natural. DIRECTV has 2.8 million subscribers, and 70 percent of those subscribers live in small towns or rural areas. More than 500,000 called themselves farmers. All of these numbers, however, were not enough to keep Channel Earth alive. It ceased broadcasting in June 1998.

"AgDay" is a nationally syndicated daily program covering agricultural news and features, weather and commodity markets. As part of the *Farm Journal* company, "AgDay" is offered via traditional television network affiliates, not via satellite. About 455,000 farmers watch the show 10 times per month, and it is carried by 160 stations nationwide.

These two efforts in agricultural television are by no means the only television offered in agriculture. There are numerous local television efforts, small networks of radio and television stations that provide agricultural programming to the farm audience. Other national farm programs include the weekly "U.S. Farm Report" by WGN in Chicago and PBS's "Market to Market."

Television, like radio, is a place where agribusiness puts little of its advertising dollars: $65 million in 1998.

Nexus Point 7

Are agricultural communicators too close to agribusiness to be objective? Why or why not? What distance from agribusiness should be maintained by agricultural journalists? By agricultural communicators? Is there a difference? Why or why not?

Information Services

The advent of electronic and digital technologies created new opportunities for agriculture's media and information providers. One of the earliest entrants into the electronic information age in agriculture was DTN/FarmDayta, an electronic information and communication services company headquartered in Omaha, Neb. DTN began with the idea of marrying computer technology with information delivery in a way that would not "scare off" farmers, who at the time were like the rest of Americans, fairly computer illiterate. DTN devised a computer-like terminal with a screen and a simple toggle bar system, and fed information via low-level FM radio waves into homes primarily in the Midwest. Farmers subscribed to DTN, and the subscription included the computer setup for a monthly fee. The farmer simply toggled his or her way through pages of information, markets, features, news and weather that were updated frequently.

DTN's sophistication began to develop in both the quality and the sources of the information presented. In arrangements with universities, more and more state-specific research and Extension information was presented on DTN's pages.

In 1998, DTN/FarmDayta had more than 159,000 subscribers in the United States and Canada. The company has introduced on-line marketing services for farmers and has moved to a satellite network for delivery of its information.

Computer technologies have changed the way that agricultural communicators work. Information can be accessed and exchanged more readily through the Internet and electronic mail. Computers are also commonly used for desktop publishing and other production tasks.

> ***Professionals currently in the field need to be constantly retooling. If they don't, they won't last long. By retooling, I mean continuous learning, seeking out opportunities to learn new techniques, participating as a judge for contests to be exposed to the ideas and thinking of others, learning new technologies and identifying ways to use those technologies to reach the target market.***
>
> ***—Donna French Dunn,***
> ***American Agricultural Economics Association***

The Internet

The Internet, worldwide, has become one of the most popular places people go for information. Currently 200 million people use the Internet, with projections of more than one billion users by 2002. Farmers also discovered the Internet; however, as of the late 1990s, most were still relying on print and electronic media for information.

This trend has not curbed the recognition that existing farm media must look toward the Internet as a way of delivering their product in the future. Many agricultural media organizations have incorporated a "Web presence" into their portfolio offered for farmers and agribusinesses. Since 1995 *Farm Journal* has made "Farm Journal Today" available on the Internet. "Farm Journal Today" also provides links to companion publications: *Beef Today, Dairy Today, Top Producer* and *ProFarmer,* as well as "AgDay," its electronic cousin. The Web site has corporate sponsors and advertisers and uses Web-based resources for surveys, registration and on-line chat services.

High Plains Journal, Home Farm, Progressive Farmer, Feedstuffs and *Successful Farming* also have a Web presence. For some of these publications, published weekly, biweekly and/or monthly, the Internet has offered an opportunity for much greater news coverage of agriculture. Because they can now reach their readers immediately, these publications can broaden the information services they provide. When readers "register" as on-line subscribers to any of these publications, they are notified when "hot" news breaks. They also are routinely queried in surveys related to public opinion on various topics related to agricultural production or policy.

Advertisers quickly recognized that the Internet versions of farm publications offered one more opportunity to reach the specialized audience of farmers. Reviewing any of the Web sites for agricultural magazines on the Internet reveals a broad presence of advertisers of agricultural products and services.

Internet publications offer publishers numerous benefits related to reader feedback. Because technology allows the counting of "hits," or visits to a Web site, publishers can tell advertisers exactly how many people will be exposed to their advertisement. The Internet also provides an opportunity for quick feedback via e-mail "letters to the editor."

The role of the Internet as a complementary medium, or in some cases as a sole medium, is still not clear. Advertisers and publishers alike still are working through advertising rates and policies to make them cost-effective. Most report that by 1999, the Internet was not making money, and they feel

> ***Although the Internet has become a treasure-trove of easily accessible information, much of the information agricultural communications researchers need still is not easily available. . . . Many important case studies in agricultural public relations remain undone because of the lack of access to information.***
>
> ***—Ricky Telg, University of Florida***

it is an additional service they are providing for their readers. The development of electronic versions of publications has restructured the editorial team in some agricultural publications with the creation of Webmasters or editors for the electronic versions. Copy flow, photography and the capabilities for audio and video "bites" provide these entities with greater flexibility in the products they produce on the Internet vs. in paper form.

Agricultural Marketing

Traditional schools of journalism and mass communication often clearly define news-editorial, broadcasting, public relations and advertising programs. These definitions follow clear patterns of activity based on time-honored traditions, responsibilities and demands. In agricultural communications, however, there tends to be a blending of roles, especially between advertising and news-editorial with a bit of public relations thrown in. Agricultural marketing or agri-marketing is a large employer of agricultural journalism graduates today.

Major chemical, seed corn, animal breed, implement and other farm product producers use

> ***The biggest challenges for agricultural PR and advertising will be communicating to an audience that is getting fewer in number but more sophisticated with each passing year. The natural consolidation of the industry is leading to fewer, larger and more complex farm operations as well as the many different businesses and suppliers who support them. Mass communications messages will continue to become less effective as opposed to precisely targeted, value-added communications that build relationships with customers over time.***
>
> ***—Jeff Altheide, Gibbs & Soell Communications***

Colleges of agriculture and state Cooperative Extension Services produce publications to reach agricultural and other audiences. These media compete for the audience members' time and attention.

either in-house agricultural marketing services or pay for the services through an agency that also provides the service. Agencies may or may not handle other clients outside of agriculture depending on the size of the business and client base. Because many large diversified companies have both agricultural and nonagricultural divisions and products, an agency may handle all facets or simply the agricultural division of a corporation.

In 1998 farmers purchased $60 billion in goods and services. And, as stated previously, agribusiness spent $860 million in advertising and promotion. Advertising in publications and on radio and television is a small part of the budget many agribusinesses spend on product promotion. Trade shows, direct mail, point-of-sale advertising and incentive merchandise and travel are other components of the agri-marketing budget. Dealers, distributors, sales representatives and customers are all targets of these agri-marketing efforts.

As already cited, a great deal of money is spent in agricultural media in advertising products and services. Agricultural media has been criticized for conflicts between what role advertising plays in editorial decision making. There is a blurring of advertising and editorial lines that is becoming more of an issue in agricultural publications, and the line is nearly invisible in farm broadcasting.

Whether or not this is a problem for the audience of these agricultural media is debatable. Farmers are a sophisticated audience, perhaps enhanced by their broad choices in building their media mix. They are aware of the biases that can creep into the editorial side of agricultural journalism and understand the role of direct mailers to company marketing.

Nexus Point 8

How is agricultural communications similar to and different from other social sciences in agriculture, such as agricultural education, extension education, rural sociology and agricultural economics?

A Career in Agricultural Journalism

Agricultural journalism or agricultural communications as a career choice positions students for a career with broad responsibilities and opportunities. Probably one of the first things that needs to be recognized is that agriculture is not a field or profession isolated from the rest of society. Agriculture must function within society for it to succeed, and therefore someone working in agriculture must have a broad understanding of all facets of life, not just agriculture. Preparation for a career in agricultural communication should include a solid collegiate experience with course work in the arts, sciences and agriculture. Many agricultural journalism programs mix agriculture course work with journalism course work. This experience, although demanding, prepares the agricultural communicator with the expertise necessary for a future with broad demands.

Writing and editing are probably the most important skills for agricultural communicators, regardless of the area. It is not enough to be interested in agriculture; an agricultural communicator must be able to use the appropriate words and language to tell a reader about a process or procedure; describe a breed or variety; or relate other information that is important to a reader, viewer or listener. And, although writing is critical, the ability to edit, to review another's work critically, is equally important. In some settings, especially in a small agency or publication, the agricultural journalist might hold the position of editor and writer all in one.

Project management also will remain important across all areas, as will problem solving, critical thinking and listening. Coupling these skills with a broad educational base will allow an agricultural journalist to provide editorial leadership in agriculture, sorting out issues and topics critical for the agricultural community.

> ***In the truest sense, I don't think we'll ever exit the Information age—we just won't call it that anymore. It's the need for sharing information that I think drives one of the biggest ethics challenges facing agricultural communicators: How do we reconcile the public interest with the growing trend toward proprietary or private information?***
>
> ***—Robert Hays, professor emeritus, University of Illinois***

In addition to writing and editing, students with an interest in agricultural publishing, public relations, advertising, graphic design or video production will need to know the basics of visual communication, including photography, photo editing, design and basic graphic composition.

Today, students majoring in agricultural journalism need a strong background in science and should take the necessary courses if their present curriculum does not fulfill this need. The science of agriculture has become more and more public and critical to public understanding of issues surrounding food safety, water quality and pesticide contamination. Fewer people are connected to production agriculture, and the general public is concerned about risks associated with agriculture.

Most communicators also will need to have a grasp of economic concepts and issues and how they affect profitability at the farm level and in the food industry.

As we noted previously, more and more agricultural media are moving toward media products integrated with technologies. These technologies, among other things discussed in this chapter, have implications for students who will be entering the field. They will need a good understanding of audience needs and motivation, they will need to know how to use various electronic media, and they will need to understand more about the science and practices in agriculture as well as be committed to learning throughout their careers.

Professional Organizations for Agricultural Communications

Most careers have professional societies and organizations that work to support those engaged in a particular professional endeavor. These organizations can provide leadership development, professional development, continuing education opportunities, networking among professionals and even cama-

> ***When I started at the Journal in 1984, I wrote on a typewriter. Now I carry my office with me inside a laptop. While I doubt anyone will soon replace the convenience of words on paper, newspapers are finding new ways to make money on the information they gather. And that is thrusting reporters into new roles.***
>
> **—Scott Kilman, The Wall Street Journal**

raderie. Agricultural communicators are no different in that respect and have several organizations that can provide career opportunities and support for agricultural communicators:

- Agricultural Communicators in Education—a professional society for agricultural communicators working primarily in the land-grant university system, agribusiness and federal government (*www.aceweb.org*).
- Agricultural Publishers Association—an organization including the publishers of major agricultural media.
- Agricultural Relations Council—an organization of professionals representing trade associations, agribusiness and private individuals interested in promoting agriculture. This organization is now affiliated with the National Agri-Marketing Association.
- American Agricultural Editors' Association—the professional organization for editors, writers, photographers and other communication professionals serving the nation's farm magazines (*www.ageditors.com*).
- Cooperative Communicators Association—an organization of professionals who communicate for cooperatives (*www.coopcom.com*).
- International Federation of Agricultural Journalists—a forum for 5,000 agricultural journalists in

more than 20 countries all over the world supporting the practice of journalism according to the principles of freedom of the press (*www.ifaj.org*).

- Livestock Publications Council—the organization that includes representation from each of the publications dedicated to breed associations or livestock trade.
- National Agri-Marketing Association—a professional development organization for agrimarketing professionals that is professionally affiliated with the Agricultural Relations Council (*www.nama.org*).
- National Association of Agricultural Journalists —an organization of agricultural journalists, magazine editors and writers that emphasizes the importance of agricultural news.
- National Association of Farm Broadcasters—the professional society for farm broadcasters—both radio and television (*www.nafb.org*).

Some of these professional societies have student chapters or memberships or student and mentor programs in which professionals and students can begin to network to create a foundation that might lead to employment upon graduation.

4

The Age of Discovery

Research in Agricultural Communications

Agricultural communications has a long history, as discussed in Chapter 1. Farm publications thrived in the early 1800s because of the agricultural knowledge and practical editorial skills possessed by early agricultural editors. Interestingly, the field was well developed before the 1862 Morrill Act, which established the first land-grant universities in the United States.

During this era, much of what was known about agricultural communications—the experience and skills needed to do the job—was learned by working editors and writers and passed down to newcomers through apprenticeships and other types of on-the-job experience. In fact, one of the reasons that early agricultural publications went out of business in the 1800s was the lack of successors. In other words, knowledge in agricultural communications and journalism resided mostly in individuals and not in any central place or form that could be easily accessed by those seeking information.

All that is known about the field today still is not collected in a single place or even several places. Yet, it is possible to access from nearly any library a growing number of books, magazines, magazine articles, videotapes, industry newsletters, journal articles and many other educational media on topics in agricultural communications and journalism.

Individuals who have access to the Internet can find many of these same sources as well as hundreds of relevant Web sites from their home or business computer. The collective body of published information on agricultural communications topics—whether in print, electronic, video or other formats—comprises what we refer to in this chapter as the literature of agricultural communications.

Body of Literature

The literature of agricultural communications, as with most other disciplines, results from the efforts of many different individuals and organizations. These individuals include government and university researchers as well as communication practitioners sharing their collective knowledge through mass media, industry publications, Web sites, graduate theses and dissertations, academic and peer-reviewed journals and other specialized outlets. Other important players include agribusinesses, media organizations and professional organizations in agricultural communications that sponsor or otherwise support research efforts.

Nexus Point 9

Who should determine the content and priorities of university teaching and research programs? Scholars? Practitioners? A mix of the two groups?

Of special concern in the current discussion is the portion of literature that has been developed through research. One of the first things students and professionals will notice about this literature is its eclectic nature. The diversity in our literature's content, purpose and methodology reflects the important fact that agricultural communications is a complex social phenomenon involving individual and group perceptions and behaviors. As such, it is amenable to study from a range of social science disciplines, especially social psychology and sociology.

When the methods and theories used in other disciplines are applied to the study of agricultural audiences or information systems, we typically claim this work as agricultural communications research. At the same time, we must note that agricultural communications research is part of a much larger body of social science literature, in which the boundaries among disciplines are often not clear-cut. As scholars and practitioners from other disciplines bring their unique theories and assumptions to the study of communication, they sometimes spawn new lines of critical inquiry that further diversify our literature.

The types of communication research conducted in agricultural communications run the gamut from simple descriptive studies conducted by one or two people to large national or longitudinal studies that require sophisticated scientific methods and teams of researchers. The diversity of research found in the literature creates a dilemma for newcomers because it is not possible to provide a quick or simple discussion that will adequately summarize its content. However, it is possible to identify several of the broad areas of investigation that characterize the field today. One way to approach the variety of work is to consider how research addresses the four components of the classic SMCR communication model[1] developed by David Berlo.

According to the model, the communication process involves four basic components in the following relationship: A *source* communicates a *message* over a *channel* to a *receiver.* As shown in Figure 4.1, these four components can be used to classify and describe some of the major areas of study in agricultural communications, depending on the primary emphasis and objectives of the research. It turns out that the categories are not mutually exclusive, because some types of research could arguably be classified within two or more of the areas. Collectively, however, the four components form a useful framework for introducing a large and growing body of research.

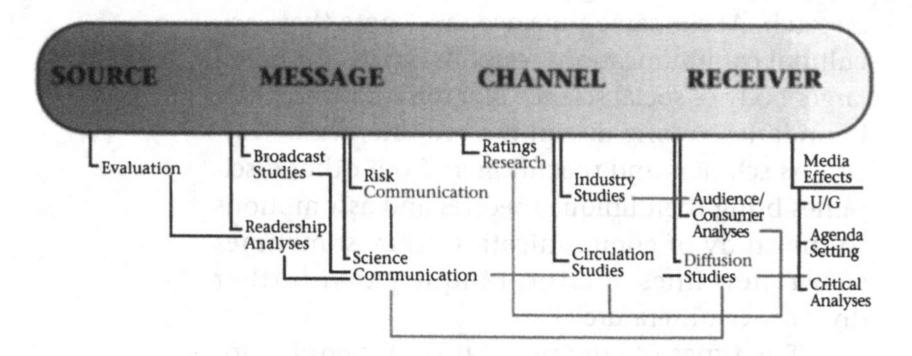

Figure 4.1. *Selected bodies of research in agricultural communications. The shaded background indicates that boundaries are not clearly defined among the four components of research. Note: U/G, uses and gratifications.*

A distinct advantage afforded by the use of the SMCR model is the recognition of a wide array of sources often using multiple channels in their competition for target audiences. It is understandable how a diverse literature has developed in agricultural communications when we consider the opportunities for research in today's complex media environment. Let us now examine each of the four components in further detail to see how research has enhanced our knowledge of each.

Source

According to the SMCR model, the source serves as the point of origin for information carried by interpersonal or mass media channels. Either individuals or organizations may be information sources, depending on their role in the communication process. Whereas the individual is typically the unit of analysis in interpersonal communication research, the organization is the more frequent unit of study in mass communications research, including agricultural communications.

Some students may have difficulty viewing organizations as sources of information. Organizations, after all, are inanimate entities that are incapable of communicating except through individuals. However, note that the influence attributed to an individual can often be traced to his or her professional or occupational affiliation with an organization, company or agency. Individual communicators in these scenarios often are acting primarily as spokespersons or representatives for their companies or organizations, not as individuals representing their own private interests. In such cases, it is more appropriate to view the organization—not the individual—as the true "source" of information.

Agricultural communications research will become more issues-oriented over the next 10 years. Researchers will focus on how agricultural industries communicate such issues as environmental conservation, waste management, chemical applications, food safety and health concerns to the public. These issues are of great importance not only to agricultural communicators, but also to agricultural industries to determine consumers' understanding of agriculture's role in these vital areas.

—Ricky Telg, University of Florida

Organizational sources in agricultural communications include commodity organizations, agribusinesses, colleges and universities and state and federal government agencies. Most of these organizations employ professional communicators or contract with communication agencies to manage their external information and/or advertising programs. While communication professionals rely heavily on their collective knowledge and experience in their editorial decision making, they often need additional information to help assess past performance and plan future programs. Many often conduct a special type of self-study called evaluation research.

Evaluation research is conducted by a group or organization to gauge the effectiveness of its current or future internal or external communication programs and activities. For instance, an agricultural communications department at a state university might sponsor a readership study to gauge the popularity and use of one of its client publications. Another example might include a corporate communications department that conducts consumer focus groups to test audience reactions to a new advertising concept. In each of these cases, the source controls the timing and content of its communication and conducts evaluation research to measure, fine-tune or otherwise improve its effectiveness. The research effort may address any aspect of an organization's communication program or audience and is considered successful if it provides needed information that was not previously available.

In addition to generating data on individuals' attitudes and preferences for specific messages or channels of information, evaluation research may address audiences' perceptions of an organization's larger public image. Image studies provide valuable information about how individuals view an organization or industry and often identify specific positive or negative attitudes that influence these judgments. Such findings have direct implications for planning or modifying a company's public relations and informational efforts.

Characteristics that may be examined in image studies include the organization's perceived credibility, relevance, commitment to ethical standards and commitment to community or industry service. Organizations that score high on these and related factors enjoy a positive public image that may be further solidified and reinforced through their informational and public relations programs. Organizations that score low on any or all of these factors are likely to adjust or otherwise intensify their public relations efforts, although communications alone is unlikely to solve the image problem.

Media organizations have an especially critical stake in building and maintaining an image of objectivity and high ethical standards. Perceived conflicts of interest, unbalanced reporting and infringements on personal privacy arc a few of the issues that have faced newspapers, radio and television news departments in recent years. As discussed in Chapter 5, the appearance of an ethical lapse can have a devastating effect on an organization's image, and profits, regardless of whether the perception is factual. These difficulties put an extra burden on organizations to protect their public image and make evaluation research an especially important tool in this effort.

Evaluation research can serve many purposes for the organization that sponsors it, but it may have value to those outside the organization too. For example, readership studies (discussed subsequently) often provide useful information about an audience's communication habits, editorial preferences and demographic characteristics. Such information can be used for a wide variety of purposes by any number of people. Likewise, research findings from focus groups can provide detailed audience information that would be difficult or impossible for a single individual to gather through other sources.

However, note that there are at least two drawbacks to attempting to use another organization's evaluation data for one's own purposes. One drawback is that evaluation research is often conducted for a highly specific purpose, so its findings cannot always be generalized to other situations. A second potential drawback is the problem of access. Because evaluation research is conducted by an organization usually for in-house or private use, findings are frequently not published or made available to outside users. Access can be particularly difficult if the research was sponsored by a commercial organization as opposed to a government agency. As you will see later, the type of sponsorship—commercial or noncommercial—has important implications for those assessing and using different types of research in agricultural communications.

Message

Most of the effort expended in communication has to do with creating and transmitting a message. Mass media messages, including those contained in advertisements, news stories and radio or television broadcasts, are the culmination of work from many different professionals involved at various stages. These communication products are doomed to failure if they are not both relevant and understandable to their target audiences. The time, energy and expense required in professional communications suggest compelling reasons for research to help streamline and improve the process.

Included in the category of message research are studies that examine various characteristics of messages themselves and how these messages can be modified to match the needs and preferences of various audiences. An example is a broad group of studies that comprises the area of readership research. Although readership studies vary widely according to their application, their common purpose is to identify how different forms, or channels, of printed information—magazine articles, newsletters, special reports—can be packaged to enhance their understandability and appeal to various audiences.

Some of the traditional aspects of readership studies include perceived readability or difficulty of text; responses to variations in typography, layout and design; and influence of photographs and other graphics on comprehension of the subject matter. The reader-friendly appearance of many contemporary newspapers, magazines and annual reports is based on the knowledge gained through years of readership research.

Radio and television stations use different types of message research to match their programming to their audience. Focus groups, personal interviews and mail and telephone survey research are some of the common methods used to gauge audience preferences for program selections, format and scheduling of news and entertainment.

Readership and broadcast audience analyses are particularly important in agricultural communications. Agricultural information often is technical and filled with the jargon of the agricultural sciences. The challenge of disseminating agricultural and technical information to lay audiences has faced public- and private-sector communicators throughout the 1900s. In the 1950s agricultural communicators made a concerted effort to respond to this problem by establishing a successful program known as the National Project in Agricultural Communications (NPAC).[2] NPAC funded a number of research-based workshops and clinics aimed at applying social science research findings to agricultural communications and encouraging additional research in such areas as readability and visual communication.

Whereas farm and rural audiences were an early focus of message research, nonfarm consumers have become increasingly important target groups for science news, including many types of agricultural and environmental information. Public awareness of science news is important because consumers need and want research-based information to avert risks and make basic decisions about their health and welfare in modern society. Food safety, water quality and household radon are just a few of the consumer risk topics that have received major media coverage in recent years. Communicating with lay audiences about these and other hazards has led to research and study in a new area known simply as risk communication. A basic goal of risk communication research is to study how best to convey scientific and technical information, including its complexities and uncertainties, to consumers who may lack a basic understanding of scientific concepts.

Another important reason for increasing public awareness is that the scientific community relies on public funding for much of its research. To support such research, consumers need to be aware of its basic purposes and possible benefits. Many land-grant universities, for instance, produce magazines, annual reports and television programs to inform alumni, legislators, community leaders and others about their successful research and extension programs.

A large and established body of message research—much of it in journalism and communication—has developed to aid the mass media in attracting and maintaining consumer interest in science topics. Research has examined audience preferences for different presentational styles and content of science topics, as well as the factors that influence these preferences.

One of the recurrent findings of these studies is that interest in the subject matter often far outweighs other factors, such as method of presentation, in predicting which science stories will be read, watched or listened to in the mass media. Individu-

Data analysis tasks that used to require skilled programmers and mainframe computers can now be handled quickly and easily with desktop computers and software.

als in a target audience have been shown to attend to particular messages for different reasons, including genuine curiosity or interest in the topic itself, or perhaps to gain information that will help them save money or perform a task more safely. Economic motives have proved to be especially important to farm audiences.

Channel

According to the SMCR model of communication, the channel is the vehicle used to deliver a message to a receiver. In the mass communications context, the channel might be a newspaper, magazine, radio or television program or other medium. Decades of research have established that individuals tend to compare communication channels according to different criteria, and these comparisons form the basis for their decisions to use and value certain media channels over others for specific types of information. Channel research is often conducted by organizations to improve their communication programs. Another type of channel research is often

Finding Research Topics

Research topics often follow broader societal trends. Jim Evans describes some of the particularly important milestones in agricultural communications:

I believe that many changes in agricultural communications mirror developments in the broader arenas of media, information systems and agriculture. Here are several agricultural communications changes of the past 100 years that seem especially notable:

- ***New and changing information technologies, especially electronic. Examples: radio, television and other audiovisual media, computer-based and interactive electronic information systems.***
- ***Changes in the financial base for the commercial farm press, as expressed in relatively greater reliance on revenue from advertisers and less reliance on revenue from readers.***
- ***More special-interest communicating within agriculture. Examples: more communications efforts by agricultural organizations and more public relations efforts by agricultural marketers.***
- ***More rural-urban communicating about issues of shared interest. Examples: environmental quality, land use, food supply and safety.***
- ***Greater segmentation of audiences for agriculture-related communications, reflecting increased specialization and concentration within production agriculture.***
- ***Greater concentration in ownership of communications media and in ownership within the agricultural marketing sector on which the media rely, financially.***

—Jim Evans, professor emeritus, University of Illinois

used to assess the performance of a group of channels, such as daily newspapers or farm radio stations in a particular area or of a particular size. Among the most important used in the communications industry are ratings research in television and radio, and various types of circulation research conducted for print media.

Ratings research refers to a variety of studies conducted by independent companies to help document audience size and composition for television and radio stations. Broadcast ratings are used extensively by advertisers in determining what mix of media will help them reach key audiences at selected times for the least cost. Although ratings research provides only estimates of audience size, it is critical to the success of television and radio stations because of its wide acceptance and use by advertisers. Circulation research is often conducted by organizations known as audit bureaus to verify the accuracy of audience and subscription data that print publications use to attract advertisers. After an audit bureau has been paid to verify a publication's circulation, the bureau's logo is often featured in the publication's masthead to alert advertisers that audience data have been validated by an independent source. Most ratings and circulation research are proprietary, meaning that it is privately owned by a company or organization that may or may not make it available to outside users.

Whereas ratings and circulation research are intended to describe audiences, other types of channel research seek to explain them. This is an ambitious goal because of the inherent difficulties in predicting human behaviors. Communication behaviors have been shown to be particularly complex and idiosyncratic. Editors, program managers, advertisers and marketers have long puzzled over the factors that influence consumers' reading habits, preferences for radio and television programs and other communication behaviors. Attempts to understand consumers' communication behaviors have led to literally thousands of research studies in recent years.

Within agricultural communications, agribusiness companies are major supporters of such research, along with media organizations and professional communications groups. An example of the latter is the National Association of Farm Broadcasters, which periodically supports research that tracks how farmers use radio and television for different types of agricultural information. Another example may be found in the periodic studies that track agricultural advertising expenditures in farm magazines and other media. These studies frequently include information on advertising expenditures by categories such as animal health or crop chemicals as well as the amount spent per advertisement.

The volume and diversity of literature in agricultural communications cannot be captured in a single textbook. Students are encouraged to review some of the publications listed in Table 4.1 to pursue specific research areas of interest. General interest magazines, newspapers and Internet Web sites may also periodically carry relevant information.

These industry studies provide valuable decision data to various sectors of the mass communications industry, whether their purpose is to track advertising trends, provide detailed demographic data about target audiences or any other number of issues. Many professional communicators and agricultural marketers use these studies as a source of secondary data, in addition to studies they may conduct on their own.

Government agencies and universities have also made important contributions to channel research. Universities have been most influential in helping create a cumulative body of rigorous com-

Table 4.1 Sources of agricultural communications literature

Industry Publications

Advertising Age
AgriMarketing Magazine
American Demographics
Billboard
Columbia Journalism Review
Editor & Publisher
Folio
Magazine Design and Production
Quill
Washington Journalism Review

Academic and Peer-Reviewed Publications

ACE Quarterly (see *Journal of Applied Communications)*
Agricultural History
Agriculture and Human Values
American Journalism
American Journal of Psychology
American Journal of Sociology
American Sociological Review
Annual Review of Sociology
Communication
Communication Education
Communication Monographs
Communication Quarterly
Communication Research
Communications: The European Journal of Communication
Communications and the Law
Communicator of Science and Technical Information
Critical Studies in Mass Communication
Design Issues
Graphic Arts Monthly
Human Communication Research
Journal of Advertising
Journal of Advertising Research
Journal of Applied Communications (formerly *ACE Quarterly)*
Journal of Applied Psychology
Journal of Broadcasting & Electronic Media
Journal of Communication
Journal of Communication Inquiry
Journal of Consumer Research
Journal of Extension
Journal of General Psychology
Journal of Marketing Research
Journal of Mass Media Ethics
Journal of Media Economics
Journal of Newspaper and Periodical History
Journal of Technical Writing and Communication
Journal of the Community Development Society
Journalism and Mass Communication Quarterly
Journalism Educator
Journalism History
Journalism Monographs
Mass Communications Review
Media, Culture and Society
Media History Digest
Newspaper Research Journal
Psychological Reports
Public Opinion Quarterly
Public Relations Review
Quarterly Journal of Speech
Rural Sociologist
Rural Sociology
Social Forces
Sociological Quarterly
Sociology
Sociology of Rural Life
Social Science Quarterly
Telecommunications Policy
Written Communication

munication research for use by educators, industry and thousands of media and communication organizations. One of their key contributions is in developing innovative new methods for conducting channel and other types of communication research and in perfecting the measurement of key communication concepts. Because colleges and universities have a public education mission and many are publicly funded, much of their research has been published in peer-reviewed journals and other outlets that are readily available to students, professionals and other researchers. (See Table 4.1 for a partial list of these resources.)

Another of the unique contributions of university research is its inclusion of a range of social scientists with interests in the study of communication. Because news media are major outlets for delivering information to the public in democratic society, it is not surprising that both channel and receiver research have attracted the interest of social scientists outside of communication and journalism. Some of the major bodies of receiver research are introduced in the following section.

Receiver

The receiver in the SMCR model refers to an individual within an audience targeted by the source. Among the earliest types of receiver research conducted specifically in agricultural communications were studies that assessed how farmers received information and how land-grant colleges of agriculture and Extension services could better serve their information needs through publications, field days and other means. These studies continued as new media came into widespread use—radio in the 1920s, television in the 1950s, and computers and related electronic technologies beginning in the 1970s. Certainly such studies have a strong "channel" component as well, but here we address that work as receiver, or audience, research because its ultimate goal is to provide a clearer picture of individuals' media preferences in combination with other personal data. The rapid development and adoption of new information technologies ensure continued research in this area.

Still other types of receiver research are geared toward helping organizations and media operate more efficiently in the competitive marketplace. The variety of work can be referenced under a number of different headings in the literature. For instance, we may speak of audience research in the journalism context, lifestyle research in advertising and consumer behavior research in marketing. Although there are differences in purpose and execution among these research areas, all seek to generate specific types of information about a group of people—a specific audience or target market. The information typically has both social and psychological variables, including conventional demographic data (age, gender, income), in addition to items that measure individuals' perceptions or awareness of any number of issues related to the communication behaviors under consideration.

Information gained through receiver research is commonly used to help communicators identify specific subgroups, or segments, within the larger audience to which messages may be targeted. Such research overlaps considerably with a specialized body of work known as diffusion research. "Diffusion" is a unique type of communication in that the message usually introduces an idea or innovation that is new to the receiver. One of the key elements of diffusion research is its use of "adopter categories" to describe and classify individuals in a social system with respect to their innovativeness, or likelihood of adopting a new idea earlier than others. In one of the widely used schemes,[3] individuals are classified into one of five categories ranging from risk-taking innovators, who usually are the first to adopt a particular idea or innovation, to laggards, who tend to be suspicious of and resistant to change.

Researchers in both the public and private sectors have found diffusion research invaluable for studying the complex processes by which individuals decide to adopt or reject various innovations. Communicators, for instance, have adapted the research to study how news and other media messages are disseminated to and among individuals and subgroups in a population, whereas marketers have borrowed its techniques to study the decision-making process used by different consumer groups in adopting new commercial products and services. The research has found particularly wide application in agriculture, because social scientists have studied the behavioral processes of large and small farmers in adopting the use of such innovations as hybrid seed, conservation practices and computers.

As one would expect, receiver research has many purposes, and all are not necessarily conducted with the primary intent of helping communicators to do their jobs better. Many of the studies, instead, are aimed at helping researchers understand how editors and reporters make decisions about what issues are covered by the news media, the factors that influence editorial decision making and the impact or effects of media messages on their audiences. An extremely rich body of research into media effects has developed in response to the notion that pervasive mass media messages may have unintended or otherwise negative effects on some individuals, particularly children and young adults. Scholars have pondered related research issues from fields other than communications and journalism for several decades.

One of the established areas of media-effects study is that of uses and gratifications research. This research is aimed at discovering the factors that motivate individuals to choose some media programming and channels over others. In this theoretical approach, individuals in an audience are viewed as autonomous and proactive consumers. As such, they are thought to make conscious decisions about the types of media and programming they select, and to base selections for news and entertainment on their own perceived needs and preferences. Research has attempted to identify the various criteria used by individuals in making these decisions, as well as their assumptions and motivations at work in forming preferences for certain media and programs. The usual strategy is to test whether the social and psychological characteristics of individuals might be used to predict their media preferences and behaviors.

Not all researchers believe that studying uses and gratifications offers a valid explanation of individuals' communication behaviors. Some of the alternative theoretical approaches conceptualize audience members as being relatively passive receivers of information who are faced with a limited range of programming choices. Within this context, the source is argued to wield relatively more power than the receiver in the communication process.

More recent work in media-effects research includes gatekeeping or agenda-setting studies. These metaphors refer to the notion that mass media reporters and editors serve an information "filtering" function by selecting what items are and are not reported to the public. The broad objective of these studies is to describe the nature of media influence on the individual. This research usually has two phases: (1) to measure and categorize media content for a specific medium or media in a given area, and (2) to measure audience members' awareness and perceived importance of issues in the news. The extent of correspondence, or likeness, between media content and individuals' perceived importance of that content indicates the degree to which media have "set the agenda" for specific public issues. Agenda-setting studies may also focus on intermedia influence, such as how content and coverage in some media are emulated by other media for certain issues and events. Content analysis and survey research are the common methods used in this research.

A growing body of social science research approaches agenda-setting from the perspective that decisions about news coverage are made with the intent of serving specific social groups or corporate interests at the expense of others. For instance, recent media studies and related research by sociologists view news organizations not simply as objective carriers of information, but as social institutions with their own norms, cultures and objectives, some of which may influence news content. Findings from this type of research often attempt to describe message content as being associated with or influenced primarily by the traditions and characteristics of news professionals and organizations themselves.

We, as professors, shortchange our students by not making them think. We lecture to them, and provide information, but rarely do we give them the opportunity to do research. Research and analysis prepare students to face the "outside world" better prepared. By teaching them how to be good researchers, we enable them to think more critically about issues and to analyze and synthesize various information sources more effectively.

—Ricky Telg, University of Florida

Researchers working within this critical perspective argue that newsworthy individuals and events in society may be intentionally or unintentionally neglected or ignored for different reasons. Media channels may divert attention away from certain events and toward topics or individuals that present a certain view of the world that favors their interests.

Critical approaches used to study the effects of media have been debated widely among communication researchers and practitioners. Despite some of the criticisms, this work represents a new and potentially revolutionary direction of mass communications research. Although there are many ways in which similar research programs could be applied to agricultural communications channels and audiences, very little of such work has been done to date.

Evaluating Research in Agricultural Communications

Although the preceding discussion was intended to introduce some of the more common types of research in agricultural communications today, no mention was made of differences in quality or how to evaluate different types of research. In reality, all social science research, including that in agricultural communications, is susceptible to certain types of problems that can threaten its accuracy or validity. Although proper research design minimizes such threats, it is impossible to remove them entirely.[4] Variations in quality of research often can be traced directly to a study's intended purpose. For instance, tight deadlines and budgets often require evaluation research to be done as quickly and inexpensively as possible. Some of these studies were never intended to provide highly specific information for their users. If the goal is only to generate baseline data for discussion purposes, even a modest project can yield useful results with minimal effort and expense. Such studies contrast sharply with large-budget projects that use elaborate research methodologies and employ several researchers. The latter types of projects are most common when there is a need for more precise, detailed data that will be used for highly specific purposes.

While the preceding two examples may represent extremes in quality, countless other studies lie somewhere in between. Clearly, it is important to exercise caution in interpreting and making a comparison among different types of studies conducted for different purposes. Unfortunately, there is not a single checklist one can use to ensure a basic level of quality for all studies. The indicators typically used to evaluate research vary by a study's design and require technical knowledge to use properly. With a little extra time and effort, however, it is possible for anyone to make some basic judgments about the quality and usefulness of a given study. When encountering studies in the literature, users should always consider the following two questions: (1) How was the research conducted? and (2) Who sponsored the research? Each is discussed in detail next.

1. How Was the Research Conducted?

A common goal in research is to generate data for planning, decision-making or educational purposes. But data are only as good as the methods used to obtain them. Using an inappropriate methodology or trying to cut corners at crucial points in a study can lead to misleading or otherwise invalid findings. Users must try to determine whether proper research methods were used before deciding whether or how to use a study's findings for their own purposes. In the case of agricultural communications research, users are likely to encounter two broad types of studies in the literature: quantitative and qualitative.

The most common research design using quantitative methods is that of mail or telephone survey research. Survey research is a common task in both communications and marketing, particularly when there is a strong need to assess attitudes, preferences, interests or demographic characteristics of a large group of people. Regardless of the purpose of the research, users should examine several specific aspects or details of these studies before using the findings. Here are a few questions that should be considered:

- Was the questionnaire field tested and assessed for validity before use?
- Was statistical or other evidence provided to ensure reliability of the questions used in the study?
- If sampling was used, was the sample of sufficient size and was the sample assignment conducted randomly or using the proper selection criteria?
- Was the response rate—calculated by dividing the number of actual respondents by the size of the sample or population—of sufficient size to generalize to the sample or population?
- Were the proper statistical tests used and reported in analyzing the data?

Answers to these questions provide only basic information about a given study. The real value in repeating this exercise is that users eventually will begin to see patterns in how research is conducted and presented, especially across different publications and topics. Recognition of those differences provides a basis for assessing and comparing the quality of different studies.

The same is true when reading qualitative research, which has grown in popularity in recent years. Used correctly, qualitative methods such as the use of focus groups and in-depth interviews can provide a wealth of rich data that would be difficult or

Focus groups, one qualitative data collection technique, are used in both industry and academia. Focus groups represent an ideal setting for guided discussion by a knowledgeable group.

impossible to gather from other methods. One of the major uses of qualitative data is to provide information to help plan future studies, including survey research. Those wanting to use focus-group findings can get some sense of a group's value or quality by asking several basic questions:

- Was the question route tested before the interview or focus group was conducted?
- Did the moderator involve all participants throughout the discussion?
- Were a sufficient number of focus groups conducted to ensure adequate comparison of groups? Were a sufficient number of people involved in the groups? How were they selected? Are they adequately described?
- Does the summary report provide sufficient detail or a transcript to allow users to interpret findings for themselves?

Many excellent books and other publications are available to those interested in learning more about specific research methods and techniques. Your instructor can provide more information and recommendations about such resources.

2. Who Sponsored the Research?

In addition to noting certain details about a study's methodology, users of agricultural communications research should be sensitive to what groups or individuals have financed studies they are interested in using. It is particularly important to distinguish between commercially and noncommercially sponsored research found in the literature.

Commercially sponsored research includes all studies, regardless of type or focus, that are funded by private companies or organizations. Within agricultural communications, agribusiness companies and communication agencies (public relations and advertising) are major commercial sponsors of research. This category might also include the research conducted by professional communication groups, such as the Agricultural Publishers Association and the National Association of Farm Broadcasters, because most of their members are employed by commercial businesses in agricultural communications.

The point to keep in mind when evaluating commercially sponsored studies is that communication professionals have particular needs in research, and their usual purpose is to generate information for highly specific purposes. Students and professionals should thoroughly understand the findings and their limitations before generalizing them to other situations.

In addition, it is possible to find industry studies that are conducted with an implicit purpose to put a particular company or aspect of an industry in the best possible light. Such studies are not necessarily faulty, but they can be misleading if used for general information purposes. The best strategy

Ethical Dilemma

When students enter the workforce, some of the most confounding issues they encounter deal with ethical choices. What are the ethical demands of the job? What are the ramifications of those choices? Like all other aspects of communications, research involves ethical choices, and agricultural communicators are not exempt from these issues, as the following guide points for discussion illustrate.

This scenario was provided by Robert Hays.

1. *A large biotechnology firm offers several million dollars to sponsor research at a public university but demands proprietary interest in the findings. Should the university agree to this? Why or why not?*
2. *The university agrees and accepts the funding. Scientists doing the research, working in university laboratories and collecting salaries, paid in large part by the taxpayers, develop a new seed variety and chemical herbicide combination that can quadruple grain production at little increased cost. The scientists want to share at least the basic principles on which their development is based through scientific journal articles; the biotech firm says they can't. Is this fair? Why or why not?*
3. *Select one of the following roles: (a) a public relations professional employed by the biotech firm, (b) a public relations professional working for the university, (c) an account executive working for an advertising agency that handles advertising for the biotech firm, (d) a farm broadcaster, farm magazine editor or newspaper farm editor working for an independent commercial medium. Take a position on this issue and be prepared to explain and defend it both as a recommendation to your employer and as a public stand.*

—Robert Hays, professor emeritus,
University of Illinois

when using commercially funded research is to request the full report of the study from the sponsoring organization and approach the data and findings with caution. When possible, users should compare findings with similar studies to confirm their validity and broaden their application.

Noncommercially sponsored research includes an extremely wide variety of studies funded by nonprofit foundations, government agencies and colleges and universities. These organizations fund and conduct research for their own informational needs, to support an educational or academic mission or for other purposes. Unlike industry sponsors, they usually have no direct link to the medium or industry being studied and no direct profit motive for conducting the research.[5]

Much of the communication research and theory you have studied in college arose from research sponsored by noncommercial sources. Countless communication models, theories and studies have been developed by university and other nonindustry researchers in the past several decades, including the SMCR model of communication used to guide the discussion of research in this chapter.

Students and professionals are likely to notice a major difference between the writing styles of commercially and noncommercially sponsored research reports. The latter are usually presented in more formal language and have a more theoretical tone than most industry-sponsored reports because they are written primarily for other communication researchers and social scientists, and not communication practitioners. On the plus side, noncommercially sponsored research reports are more likely to contain detailed information about the methodology and techniques used to analyze the data, as well as additional information about similar studies and their findings.

Conclusion

Improving ourselves as professional communicators means looking for better ways to do things. Research is one of the most valuable tools we have to improve our performance. As we have discussed, even those without the time or budget to conduct their own research can benefit from the variety of commercial and noncommercial studies published in the agricultural communications literature. To glean the full benefit from this literature, users need to develop a basic understanding of social science research methods.

Nexus Point 10

What reasons can you cite for the lack of graduate programs in agricultural communications and agricultural journalism? What implications might this have on the profession of agricultural communications and agricultural journalism?

There are a variety of ways to acquire the necessary skills. One is to learn directly from other professional communicators who conduct research. Students likely will have access to such individuals during internships and employment. Participation in professional communication organizations is another way to meet and learn from those involved in communication research. Most would be pleased to share their research with you and to provide advice on how you could learn more or perhaps even get involved with a research project.

Another option is to consider taking additional undergraduate or graduate course work in social science research methods and theory. Such course work is usually available through a variety of departments, including communication, journalism and sociology. The research skills and knowledge you gain from these experiences will help you become a smarter consumer of communication research, an essential trait in the modern communication age.

Notes

1. For more information on the SMCR (source-message-channel-receiver) model, see Berlo, D.K. (1960). "The Process of Communication: An Introduction to Theory and Practice." New York: Henry Holt & Company.

2. For more information on the NPAC, see Miller, M.E. (1995). AAACE changed our lives: NPAC and agricultural communications in the '50s. *Journal of Applied Communications 79*(3), 1–9.

3. The five categories are innovators, early adopters, early majority, late majority and laggards. For a complete discussion, see Rogers, E.M. (1983). "Diffusion of Innovations." New York: Free Press. Many other schemes have been proposed to classify adopter categories. The discussion of adopter categories in this text is intended only to be illustrative of a general process used to define audiences, not to advocate the use of this specific scheme in communication research.

4. These threats are not unique to communication research. For instance, it is common practice in social science research to use individuals' perceptions as predictors of actual behavior, despite the frequent lack of correspondence among them. Other problems include difficulties in attitudinal measurement and the inability to establish causation in social research. Consult a modern social science research methods textbook for more information on these and other problems as well as suggested measures to reduce their effects.

5. One notable exception is the case in which university researchers receive grants or other funding from private industry to conduct research. In this case, university research would be considered commercially sponsored.

5

Nexus of Ideas

A nexus is a causal link. Throughout this book, we have provided Nexus Points. We hope that you have thought about some of these ideas, although they may be quite new to you. Through years of discussions and considerations of scholarly work, we have been thinking about these Nexus Points too. This chapter provides our thoughts. We take each point as it has been presented in the book and discuss it. Our discussion should not end yours; if anything, we hope that it makes you more involved with these topics.

Another aim of this chapter is to capture ideas that have not been published but are important to our field. For example, unpublished thoughts from academic leaders illuminate the basic nature of agricultural communications. We also hope that these discussions spur further scholarly efforts.

Nexus Point 1

Some academic programs are called agricultural journalism whereas others are called agricultural communications. Is there a difference? Outside of academia, is there a difference?

Agricultural communications and agricultural journalism programs are similar in academia. Usually the name of the program relates to the date it

> ***I am fulfilled beyond measure by knowing I and my publications are making a positive difference in the lives, families and businesses of my readers.***
>
> ***—Loren Kruse,* Successful Farming**

was established. Just as most academic units in communications used the name *journalism* prior to 1970, older programs in our field were named *agricultural journalism*. Programs introduced later were called *agricultural communications,* following the pattern of the parent field.

The functions of journalism and communications, however, are different. *Journalism* refers to reporting and editing for journals, newspapers and broadcast media. *Communication,* a broader term, includes entertainment, information, persuasion and advocacy.[1]

> ***If you want to learn about people, talk to people, communicate with people, dive into an industry and become committed to it, agricultural communications is a great place to be. You work closely with people who are true professionals, whether those are co-workers, farmers, clients. However, I think agricultural communications takes a commitment to the agricultural industry first. That makes you sincere in your efforts and ultimately trustworthy.***
>
> ***—Donna French Dunn, American Agricultural Economics Association***

Outside of academia, whether one is an agricultural journalist or agricultural communicator marks a major difference in functions and practices performed. An editor or advertising manager for a breed association magazine has a much different objective and audience than a writer covering agriculture for a daily paper. The work of a public relations consultant differs significantly from that of an editor for an agricultural magazine.

Nexus Point 2

How is agricultural communications different from mass communications? Or trade journalism? Should these programs be housed in schools of journalism and mass communication or in agricultural departments?

Agricultural communications, including agricultural journalism, is similar in many ways to mass communications. Certainly, editorial and other communications skills needed by practitioners in both fields are the same. The difference lies in the communicator's knowledge of technical subject matter and the intended audience. The agricultural communicator is expected to bring with him or her a level of specialized knowledge in the agricultural field that typically is not required of the mass communicator.

Bearing in mind that agriculture is a diverse, applied science and business, agricultural communicators may have knowledge of production agriculture, horticulture, soil science, hydrology, forest ecology, food science, milling or any of the myriad of agricultural fields. They also generally do not master all of these fields. A communicator who has studied agricultural economics and can describe risk management strategies through the use of options and futures may not correctly identify a Maine-Anjou heifer and yet still be a qualified agricultural communicator. Within their fields of expertise, agricultural communicators are expected to carry with them specific knowledge.

Recruiting qualified professionals with a mastery of both agricultural and communications subject matter has challenged agricultural media and business employers for years. In cases in which university-trained agricultural communicators were not available, employers had two options: hire an agriculturalist and provide on-the-job communications training, or hire a trained communicator who would be expected to pick up agricultural knowledge on the job. Given the dichotomy of skills required, it could reasonably be argued even today that university programs in agricultural communications could be housed in either agricultural or journalism and mass communications departments.

Agricultural communications sits in an area between mass communications and trade communications, at times sharing objectives with both but not entirely overlapping either. As with mass communications, many agricultural communications messages are of interest to mass audiences.

Like trade communications, agricultural communications often involves specific language related to the field. Also agricultural communications is concerned with the end use of the communications and actions of the receivers, not simply with the dissemination of the message.

Students entering the agricultural communications field today need good fundamental communications skills—writing skills, speaking skills, an appreciation for photographic and visual communications, etc. They need an understanding of the many communications vehicles in use in agriculture today and how they interact. A good understanding of the agricultural industry is very beneficial but is not absolutely critical if the student has an interest and ability to learn. For students planning a career in agricultural advertising or PR, a basic understanding of business management and marketing theory is also helpful.

—Jeff Altheide, Gibbs & Soell Communications

Nexus Point 3

What is the administrative home of your agricultural communications or agricultural journalism program? List two or three advantages and disadvantages of housing these programs within larger departments with other academic programs as opposed to placing them in their own departments.

University administrators often place newer or smaller academic programs within larger or more-established departments rather than create new departments for them. Doing so saves money for universities because programs can share administrative costs, space and equipment.

Such is the case with the majority of agricultural communications and agricultural journalism programs nationally. More than half are housed within departments of agricultural education; most of the remainder are housed with other disciplines. Such a combination makes sense in some regards. After all, agricultural communications shares important similarities with agricultural education, rural sociology and agricultural economics in that they are all social sciences (see Nexus Point 8).

On the other hand, it is equally important to recognize some of the long-term implications of consolidating academic areas into one unit or department. One is that smaller programs run the risk of losing their identity when placed within larger departments. Agricultural communications programs also frequently lose visibility under such arrangements, particularly if *communications* or *journalism* is not in the department name.

Another implication of combined departments is that important decisions about agricultural communications programs—courses offered, curriculum requirements, faculty hiring decisions—may be made by those with little or no formal training or experience in the discipline. Such individuals may not be equipped to deliver a rigorous agricultural communications curriculum that will prepare graduates for employment in a specialized, competitive field.

This discussion is not intended to give the impression that combined departments are always detrimental. In fact, one model that was common in past years drew much of its strength from combining academic and applied communications components. This arrangement involved housing agricultural communications teaching programs in departments dedicated to applied communications for the Cooperative Extension Service or Agricultural Experiment Station. The staffs in these departments were responsible for various communications functions and media, including publications, broadcasts, news, graphic design and photography. In some cases, staff members carried faculty rank; in others, only those with teaching and student advising responsibility were considered faculty members.

As previously noted, most agricultural communications teaching programs are no longer housed within applied communications units. For some colleges and universities, it is not possible because the state's applied communications unit is located elsewhere. In other cases, such a link is still possible because academic and applied communications units are on the same campus.

The trend toward separation of academic and applied communications programs has been detrimental to students pursuing careers in agricultural communications and agricultural journalism. A major advantage of combining academic and applied communications components is the opportunity for agricultural communications students to work in a "laboratory" environment, learning the ropes of publishing, agricultural news writing and media relations, or photography. Applied communications professionals can provide a direct network to the professional media with which they interact.

A second advantage to this joined model is the symbiotic relationship between academic and applied communications. Academicians—researchers and teachers—in agricultural communications can be influential in how applied communication is carried out. Their research perspectives and educational expertise can be valuable resources for communicators reaching specialized audiences with a specialized message. Applied communicators—writers, editors, designers, broadcasters and photographers—offer a rich field of study for academicians and can help guide the research agenda and utilize research findings. Each communications project, whether news story, video news release or photograph, is a subject for research and teaching critique.

Today's agricultural communications students, deprived of this immediate relationship between academic and applied communicators, face greater challenges in seeking out professional mentors.

Nexus Point 4

For various reasons some people do not have access to information technologies. What happens to people who cannot get information via electronic media such as the Internet or CD-ROM? As an agricultural communicator, how can you plan an information campaign that includes everyone?

Society is filled with haves and have-nots. Some people have money; others do not. Some have an education; others do not. Some have access to information via technology; others do not. As communicators, we should be concerned with access to technology and its implications.

As more and more information for public use is put on the Internet or into other digital formats (diskette or CD-ROM), there will be an issue of whether access is limited only to people who can afford it. If individuals do not own or have ready access to a computer, they cannot easily receive information. If they are in a rural location with limited telecommunications resources, they cannot physically receive this information. If they have limited educational and financial resources, they may not understand how to retrieve or pay for this information. And, finally, if they have little interest in or are afraid of technology, they will choose not to receive the information. These people become information have-nots.

These individuals are at a disadvantage to those who do have access to technology. In most cases, this disadvantage may not impact their daily lives; but if information is time-sensitive or critically important, such as in response to an emergency, these individuals are at a distinct disadvantage and possibly at risk more than others. As this gap in access to information widens, socioeconomic ramifications will continue to grow. Among those who have the least access to electronic communications are the rural poor, rural and central city minorities, young households and households headed by females. Those who are more electronically connected are of higher socioeconomic status and better educated. In other words, the people who might benefit most from the information available electronically are the least able to get it. This situation then increases the gap between the haves and have-nots.

Agricultural communicators must remember these information have-nots as they design the communication products on which they work. Not everyone will have the same access to a Web site or an e-mail group. Some people may still need to be reached through traditional communication means of print, radio or television. Direct mail may reach them more effectively. This means that communicators will need to design parallel channels in their communications, not simply rely on the medium with the most "bells and whistles."

Nexus Point 5

Agricultural issues are among the most important facing consumers today—food safety, environment, water quality, urban sprawl. But some have raised questions about the willingness of agricultural editors and news directors to address controversial issues in their media. How are controversial issues covered in the agricultural media? Should agricultural news and information be reported more like general news?

Although farmers comprise less than 2 percent of the U.S. population, agricultural issues regularly generate major news stories in the nonfarm press. Threats to food safety and water quality, for instance, are consistently ranked by consumers as among the most worrisome issues facing them today. These and many related issues are "agricultural" in that they involve farm products or the unintended byproducts or consequences of agricultural production. Given the seriousness of these issues to all consumers, it should not be surprising that metropolitan media would report them. But what about the farm press?

It is important to remember that farmers are consumers, too, and that their proximity to agricultural production activities puts them at additional risks not faced by their urban counterparts. With this in mind, one might expect significant coverage of these issues in the farm press. But at least some critics have argued that farm editors and news directors have avoided critical coverage of agricultural issues in their media—especially those with social, environmental and economic implications.

One of the reasons cited by critics for the lack of coverage is that financially strapped publishers have been forced to cut editorial staffs to remain in business. Smaller staffs often are unable to devote the time necessary for in-depth reporting on complicated issues. Others have pointed out that agribusiness advertising—particularly from crop chemical and animal health companies—is the major source of revenue for many farm publications and broadcast stations. Consequently, some believe that agricultural advertisers wield a strong influence in determining editorial content in the farm press, including the ability to block or prevent potentially damaging stories about their business or industry.

It is understandable that such charges draw angry responses from many agricultural communicators. The position taken in this textbook is that thoughtful, open discussion about the proper role and performance of agricultural media is a worthwhile enterprise for both students and professionals. The purpose is not to second-guess individuals' editorial judgment or professional ethics, but to assess the general state of agricultural journalism and the pressures that threaten the flow of important news and information to its audiences. Based on your understanding of the issues, are audience needs and interests being served?

> ***The journalists I work with come from all sorts of backgrounds: from lawyers to English teachers. I think the danger in someone only studying agricultural journalism is that person might become an advocate for agriculture. I'm not an advocate for agriculture. When I covered banking and retailing, I wasn't an advocate for those industries, either. I'm an advocate for the First Amendment, for journalism, and our readers making sense of the world.***
>
> ***—Scott Kilman,*** **The Wall Street Journal**

Nexus Point 6

What is service journalism? In what ways is it an appropriate description of agricultural journalism? In what contexts is it inappropriate? Do you feel that agricultural journalism is advocacy journalism? Why or why not? If not, should it serve such a role?

Some agricultural journalists use the term *service journalism* to describe their style of reporting and writing information for target audiences. Others, including some outside the profession, refer to agricultural journalism as a type of advocacy journalism. Although definitions of service journalism vary by context, we broadly define it here as a style of journalism that has an explicit goal of bringing about a change of activity or behavior on the part of the information receiver. In the agricultural communications context, applications of service journalism might involve encouraging farmers to adopt specific new conservation practices or marketing techniques. Other examples might include persuasive messages in magazines or electronic media aimed at encouraging the use of farm safety or other recommended practices or innovations.

Advocacy journalism is a style of journalism that places the communicator in the role as an advocate, representative or spokesperson for a particular group or cause. Although used less frequently than service journalism to describe agricultural communications, advocacy journalism also represents a departure from traditional journalistic techniques that stress objectivity and balanced reporting.

Service and advocacy journalism represent specific genres of communication that appropriately describe some forms of agricultural communications. Some communicators embrace and defend the terms, whereas others view them with various degrees of disdain or suspicion. Farm magazines and electronic media are the most likely channels to be associated with service journalism in the agricultural communications context, whereas daily newspapers are the least likely to favor any variation from traditional journalism.

Although many agricultural communicators view their craft as service journalism, agricultural journalism and service journalism are not synonymous. Farmers and consumers also require objective information and even investigative reporting offered by impartial, detached media. Discerning consumers of information are well served by making a distinction between service or advocacy journalism and traditional journalism—not because one is preferable to the other in all cases, but simply because they serve different purposes.

Nexus Point 7

Are agricultural communicators too close to agribusiness to be objective? Why or why not? What distance from agribusiness should be maintained by agricultural journalists? By agricultural communicators? Is there a difference? Why or why not?

Many agricultural journalists and communicators actively participate in various activities and events sponsored by agribusiness. Indeed, many national professional meetings of agricultural communicators are sponsored by large agricultural companies. However, many journalists also believe that it is important to maintain distance between themselves and the individuals and businesses on which they report. These journalists believe that maintaining this independence is essential for them to remain objective in their editorial decision making. Although media and other communications organizations often have policies and standards of ethics that address such issues, individual communicators often must decide whether a particular practice is acceptable or unacceptable in a given situation.

A case in point is that of corporate sponsorship.

Agricultural communicators have differed in their perceptions of the seriousness of this problem. Some are comfortable making decisions on whether to accept corporate sponsorship or subsidies on a case-by-case basis, arguing that the practice is acceptable in some instances. Others, including members of the National Association of Agricultural Journalists (NAAJ), have opposed sponsorship in principle on the grounds that it has the potential to cloud editorial decision making. NAAJ members pay for their own travel and expenses to national meetings and finance their own annual awards competition. Furthermore, some NAAJ members have been openly critical of other agricultural communications organizations that have allowed corporate sponsorship of their meetings.

But the issue of whether agricultural communicators can be biased or otherwise influenced by such subsidies may not be as important as whether there is the appearance of bias in the eyes of the audience. Media and other communications managers have long realized that the appearance of a conflict of interest can undermine the credibility of their organizations whether or not the conflict actually occurred. Once the credibility of an organization is tarnished, it is nearly impossible to restore.

Given this situation, one might expect agricultural communicators to be especially guarded about their perceived relationship with agribusiness. However, it is also important to remember that many agricultural communications media and organizations rely on agribusiness advertising as their main source of revenue. With this in mind, does keeping a distance make economic sense?

Nexus Point 8

How is agricultural communications similar to and different from other social sciences in agriculture, such as agricultural education, Extension education, rural sociology and agricultural economics?

Jim Evans, a long-time advocate for agricultural communications and a scholar in the field, reflected upon the question, What are the challenges facing agricultural communicators? His thoughtful response is included here.

I think that the real potential for professional agricultural journalists and communicators lies in pursuing several related goals:

They must prepare themselves to become more than expert assemblers and handlers of information. Certainly, they need technical skills in gathering, organizing, processing and disseminating information. They also need a command of the subject matter involved. In addition, however, they need to understand people and how people gather, organize and process information for decision making. This is a highly complex process, tailored to specific audiences, situations and purposes.

In an era of special-interest communicating, professionals need to understand the essence of communicating. The word has its roots in the spirit of communing, listening actively, sharing and interacting in a spirit of respect and trust. This is not a soft headed, impractical notion. In fact, it involves principles and practices used by negotiators and others whose profession is to communicate in conflict. And it is productive in recognizing communication as a complex human process. It uses advocacy. But it interprets advocacy as more than using voices and speaking out. It also involves advocacy as using ears to hear, minds to listen and learn, and hearts to care about those who hold differing points of view. And it puts advocacy within a more productive context—of joint problem solving. I have yet to find a more promising approach to communicating. Information, alone, is no substitute; it is raw material for the process.

> ***If professional agricultural communicators capture and follow this spirit I believe that they will continue to offer increasing leadership and service for agriculture and society.***
>
> ***—Jim Evans, professor emeritus, University of Illinois***

In recent years, several agricultural educators have noted the similarities between their discipline and agricultural journalism and agricultural communications, especially as it relates to vocational agricultural education programs.[2] Certainly these fields exhibit similarities with each other and with Extension education, rural sociology and even agricultural economics, in that they are all social sciences grounded in agriculture.[3]

Agricultural communications, rural sociology, Extension education and agricultural education draw from the academic bases of psychology and sociology. Consequently, these fields have some similar philosophies. In a practical sense, they focus more on process than subject matter content. In a broader sense, they are interested in the processing, flow, utility and effects of knowledge about agriculture.

Often faculty trained in other disciplines have been hired to teach agricultural communications. Usually this does not present a problem unless the faculty adhere to their original theoretical bases without adapting to agricultural communications; this could impair the process of educating agricultural communications students.

Communications and education are necessary for each other's accomplishment, although they are distinctly separate functions. To communicate, the source and receiver must share a common system of symbols that encodes their messages. In essence, this is our language. To have a common system of symbols, both must learn that system. Hence, to educate, we must be able to communicate.[4]

Agricultural communicators differ from agricultural educators because they work in different environments. Communicators deal with larger audiences and more formal media channels. Communications students are more interested in mass communications, interpersonal communications and the effects of communications. Knowledge of the workings of formal media is quite important for communicators, because they have less opportunity for face-to-face communications than educators and therefore have a greater potential for misunderstanding. Communicators work through mass media as well as public and private organizations. They produce informal, noncredit information and use media and information technology skills.

Agricultural educators, on the other hand, operate through school systems and produce formal, credit-based, degree-oriented education. Their techniques involve educational concepts and skills. Extension educators operate through Extension systems and provide nonformal, noncredit information and education. They need mass media skills for dealing with local media, but primarily function in the arena of continuing education.

Rural sociology also shares with and contributes important principles to agricultural communications. Among these is the basic notion that communication processes are *social* phenomena that are amenable to study using the conceptual tools and techniques of the social sciences. And, as in rural sociology, modern mass communications research borrows many of its most important tools, especially in theory, directly from sociology.

Nexus Point 9

Who should determine the content and priorities of university teaching and research programs? Scholars? Practitioners? A mix of the two groups?

Tensions between communication practitioners and scholars date back to the early 1900s, when many pondered whether university training was needed to enter the journalism field. That issue was gradually defused as journalism schools became more common and began to produce graduates in increasing demand for careers in journalism, advertising, public relations and related fields.

Practitioners and scholars do not always agree on how teaching and research programs in agricultural communications should be managed. Some practitioners believe that universities are too slow and bureaucratic to respond to industry needs, whereas university faculty argue that public education and research should serve broader needs than those voiced by industry. However, there is general agreement that a bachelor's degree is essential for entry-level positions in most communication fields today, including agricultural communications. There is relatively less agreement on how the degree program should be organized and delivered—the subjects that should be taught, the style and approach used to teach them and the qualifications needed by faculty.

Agricultural communications students have a stake in the argument because it has been their perceived performance on the job that has fueled the issue. One of the major complaints voiced by employers is that too many graduates lack the basic communications skills, such as in writing, speaking and listening, that they believe should have been acquired in college. Other issues raised by employers include students' lack of motivation and professionalism, as well as a basic misunderstanding of the skills and competencies needed by industry.

In response to these concerns, many academic communications programs, including those in agricultural communications, have established more course work in communications subjects, encouraged or required student participation in communications internships and mentor programs and established advisory boards of practitioners to help guide and improve their curricula. At the same time, many in the academic community have come to the defense of their students, programs and faculty.

The debate between practitioners and scholars, as well as disagreements within the groups, raises interesting and controversial issues about who should determine the content of university teaching and research programs. Some practitioners have argued that universities are out of touch with how things are done in the "real world" and need to focus on developing technical skills among students. Scholars often counter that such comments frequently originate from those who have limited familiarity with the academic programs they are criticizing and that education broader than simply the development of skills may better serve graduates throughout their careers. What is your perspective?

Nexus Point 10

What reasons can you cite for the lack of graduate programs in agricultural communications and agricultural journalism? What implications might this have on the profession of agricultural communications or agricultural journalism?

Whereas academic programs in agricultural communications have flourished at state universities in recent years, most offer only baccalaureate degrees. Those that offer graduate study in agricultural communications often rely heavily on other agricultural academic programs and departments, primarily agricultural education, for the bulk of their course work. The net result is a lack of graduate programs staffed by faculty trained specifically in agricultural communications.

The shortage of graduate programs has practical implications for those preparing for careers in agricultural communications. Once reserved primarily for those pursuing university teaching or research, graduate education is now required for a range of attractive positions in agricultural communications, management and research in both private and public sectors. Agricultural communications students and professionals who aspire to these positions often are faced with three choices:

1. Attempt to create one's own agricultural communications graduate program by combining courses from different academic departments.
2. Select other academic programs in which to major.
3. Forgo graduate school altogether.

None of these alternatives is ideal.

These individuals, however, are not the only ones affected by the lack of graduate programs. Because of their emphasis on research, graduate programs are one of the major mechanisms by which a field develops its knowledge base. Research projects, reports, presentations, journal articles, books, theses and dissertations are some of the scholarly by-products of graduate research, and the knowledge they contain is essential to a field's progress and development.

Several universities offer graduate programs in agricultural communications. In addition, administrators are faced with the challenge of finding and recruiting qualified faculty to teach graduate courses and conduct relevant research. If these individuals do not come from graduate programs in agricultural communications—and it is likely they will not—where will they come from? (See Nexus Point 3 for a related discussion.)

Notes

1. James F. Evans provided information that has served as the basis for much of this discussion. Evans, a remarkable scholar of agricultural communications, shared some of these ideas with one of the authors in 1991.

2. See Lockaby, J., and Vernon, J.S. (1998). Agricultural communications: what is its connection to agricultural education? *Agricultural Education Magazine 71*(3), 16–17; Birkenholz, R.J., and Craven, J. (1996). Agricultural communication—bridging the gap. *Agricultural Education Magazine 68*(9), 10–11; and Deeds, J.P., and Dorris, E.A. (1987). Agricommunication: a new opportunity. *Agricultural Education Magazine 60*(2), 15–16.

3. J. Evans, personal correspondence, Jan. 1991. J. Ross, personal correspondence, Jan. 1991. Also see Barrick, R.K. (1989). "The Discipline Called Agricultural Education." Professorial Inaugural Lecture Series. Department of Agricultural Education, The Ohio State University, Columbus.

4. Salomon, G. (1981). "Communication and Education." Beverly Hills: Sage.

Appendix

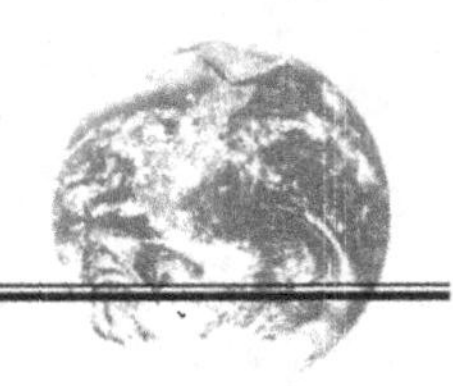

Universities with chapters of the Agricultural Communicators of Tomorrow

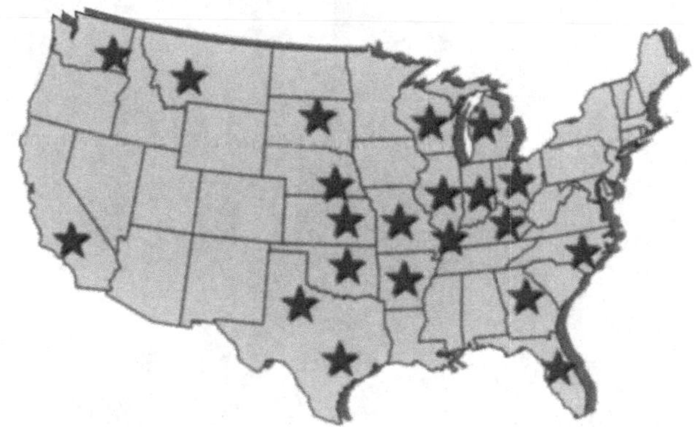

California Polytechnic State University, San Luis Obispo, Calif.
Kansas State University, Manhattan, Kan.
Michigan State University, East Lansing, Mich.
Montana State University, Bozeman, Mont.
North Carolina State University, Raleigh, N.C.
The Ohio State University, Columbus, Ohio
Oklahoma State University, Stillwater, Okla.
Purdue University, West Lafayette, Ind.
South Dakota State University, Brookings, S.D.
Texas A&M University, College Station, Texas
Texas Tech University, Lubbock, Texas
University of Arkansas, Fayetteville, Ark.
University of Florida, Gainesville, Fla.
University of Georgia, Athens, Ga.
University of Illinois, Champaign-Urbana, Ill.
University of Kentucky, Lexington, Ky.
University of Missouri, Columbia, Mo.
University of Nebraska, Lincoln, Neb.
University of Wisconsin-River Falls, River Falls, Wisc.
Washington State University, Pullman, Wash.

References

"Ag Ad Spending Tops $200 Million." (1998). *AgriMarketing 36*(2), 14.

Agricultural Communicators in Education (1996). *The communicator's handbook: Tools, techniques and technology.* Gainesville, Fla.: Maupin House.

Anderson, J.A. (1987). *Communication research: Issues and methods.* New York: McGraw-Hill.

Baker, J.C. (1981). *Farm broadcasting: The first sixty years.* Ames: Iowa State University Press.

Baker, J.C. (1985). "Farm Broadcast Blossoms in Radio's Early Days." *AgriMarketing 23*(6), 54–56.

Barclay, R.W. Jr. (1997). "The Use of Radio in Arkansas for Agricultural Information." *Journal of Applied Communications 81*(4), 41–52.

Barrick, R.K. (1989). *The discipline called agricultural education.* Professorial Inaugural Lecture Series, Department of Agricultural Education, The Ohio State University, Columbus.

Beckman, F.W., O'Brien, H.R., and Converse, B. (1927). *Technical writing of farm and home.* Ames, Iowa: Journalism Publishing Company.

Berlo, D.K. (1960). *The process of communication: An introduction to theory and practice.* New York: Henry Holt & Company.

Birkenholz, R.J., and Craven, J. (1996). "Agricultural Communication—Bridging the Gap." *Agricultural Education Magazine 68*(9), 10–11.

Boone, K.M., Paulson, C.E., and Barrick, R.K. (1993). "Graduate Education in Agricultural Communication: The Need and Role." *Journal of Applied Communications 77*(1), 16–26.

Bostian, L., and Byrne, T. (1984). "Comprehension of Styles of Science Writing." *Journalism Quarterly 61*(3), 676–78.

Brunn, S.D., and Raitz, K.B. (1978). "Regional Patterns of Farm Magazine Publication." *Economic Geography 54*(4), 277–90.

Buck, C.A., and Paulson, C.E. (1995). "Characteristics, Educational Preparation, and Membership in Professional Organizations of Agricultural Communicators." *Journal of Applied Communications 79*(2), 1–13.

Burnett, C., and Tucker, M. (1990). *Writing for agriculture: A new approach using tested ideas*. Dubuque, Iowa: Kendall/Hunt Publishing.

"Changes in Radio Ratings Research Makes Information More Accessible to Stations, Networks." (1997). *AgriMarketing 35*(10), I6–I11.

Crawford, N.A., and Rogers, C.E. (1926). *Agricultural journalism*. New York: Alfred A. Knopf.

Deeds, J.P., and Dorris, E.A. (1987). "Agricommunication: A New Opportunity." *Agricultural Education Magazine 60*(2), 15–16.

DeFleur, M.L., and Dennis, E.E. (1998). *Understanding mass communication: A liberal arts perspective*. Boston: Houghton Mifflin.

"Delivering to the TV Generation." (1997). *AgriMarketing 35*(10), I12–I13.

Demaree, A.L. (1941). *The American agricultural press: 1819–1860*. Morningside Heights, N.Y.: Columbia University Press.

Dortch, S. (1994). "Farewell to the Farm Report." *American Demographics 16*(3), 21–22. <http://www.demographics.com/publications/ad/94_ad/9403_ad/ad552.htm>.

Duncan, C.H. (1961). *Find a career in agriculture*. New York: G.P. Putnam & Sons.

Evans, J. (1972). "Broadening the Academic Base in Agricultural Communications." *ACE Quarterly 55*(4), 30–40.

Evans, J. (1989). The Importance of the Art of Communications in Agriculture. Proceedings of the 1983 Annual Meeting of the American Association of State Colleges of Agriculture and Renewable Resources, March 1984 (available from the Agricultural Communications Documentation Center, University of Illinois, Champaign).

Evans, J., and Bolick, J. (1982). "Today's Curricula in Agricultural Communications." *ACE Quarterly 65*(1), 29–38.

Evans, J.F., and Salcedo, R.F. (1974). *Communications in agriculture: The American farm press*. Ames: Iowa State University Press.

"Farm Publications on the Web: Creating Useful, Successful, Profitable On-line Services." (1997). *AgriMarketing 35*(9), 38–45.

Fliegel, F. C. (1993). *Diffusion research in rural sociology: The record and prospects for the future*. Westport, Conn.: Greenwood Press.

"Forces Shaping U.S. Agriculture—Briefing Room." (1997). *Economic Research Service, USDA* (Sept.) <http://www.econ.ag.gov/briefing/forces/index.htm>.

Fox, J.M. (1956). *Roosevelt: The lion and the fox.* New York: Konecky and Konecky.

Fusonie, A.E. (1988). "The History of the National Agricultural Library." *Agricultural History 62*(2), 189–207.

Fusonie, A.M., comp. (1975). *Heritage of American agriculture: A bibliography of pre-1860 imprints.* Beltsville, Md.: National Agricultural Library.

Gates, B., ed. (1993–1997). *Microsoft Encarta 98 Encyclopedia.* Portland, Ore.: Microsoft.

Genung, B. (1998). "NAFB Releases Its 1997 Professional Improvement Study: What Farmers Want From Farm Broadcasting." *AgriMarketing 36*(3), 50–53.

Hardt, H. (1992). *Critical communication studies: Communication, history and theory in America.* New York: Routledge.

Hartke, D. (1998). "Marketing Opportunities Revealed, Print Still King, Internet Growing." *AgriMarketing 36*(6), 34–35.

Hartke, D. (1998). "Technology Use Increasing in Agriculture, Still Shopping Locally." *AgriMarketing 36*(8), 18–19.

A History of AAEA: Roots of American Farm Magazines." (1996). *AgriMarketing 34*(9), I4–I6.

Kangas, S. (1997). "The Great Depression: Its Causes and Cure." *Liberlism Resurgent* <http://www.scruz.net/~kangaroo/THE_GREAT_DEPRESSION.htm>.

Kearl, B. (1983). "The Past and Future of Agricultural Communications. Part I: A Look at the Past." *ACE Quarterly 66*(4), 1–7.

Kessler, L. (1984). *The dissident press: Alternative journalism in American history.* Beverly Hills: Sage.

Kroupa, E., and Evans, J. (1973). "New Directions in Agricultural Communications Curricula." *ACE Quarterly 56*(3), 28–38.

Kroupa, E., and Evans, J. (1976). "Characteristics and Course Recommendations of Agricultural Communicators: An Update." *ACE Quarterly 59*(1), 23–31.

Lehnert, R. (1991). "Bitter Harvest for a Farm Magazine." *Washington Journalism Review 13*(10), 19–21.

Leonard, T.C. (1986). *The power of the press: The birth of American political reporting.* New York: Oxford University Press.

Lockaby, J., and Vernon, J.S. (1998). "Agricultural Communications: What Is Its Connection to Agricultural Education?" *Agricultural Education Magazine 71*(3), 16–17.

Logsdon, G. (1992). "Filling the White Space Between the Ads." *Agriculture and Human Values 9*(2), 54–59.

Maixner, E. (1999). "A Spirited Start." *Ohio Farmer 295*(3), 42.

Marcus, A.I. (1986). "The Ivory Silo: Farmer-Agricultural College Tensions in the 1870s and 1880s." *Agricultural History 60*(2), 2–36.

Marti, D.B. (1980). "Agricultural Journalism and the Diffusion of Knowledge: The First Half-Century in America." *Agricultural History 54*(1), 28–37.

McQuail, D. (1985). "Sociology of Mass Communication." *Annual Review of Sociology 11*, 93–111.

Miller, M. E. (1995). "AAACE Changed Our Lives: NPAC and Agricultural Communication in the '50s." *Journal of Applied Communications 79*(3), 1–9.

Murphy, J. (1997). "Custom Publications Provide Additional Tools for Marketers." *AgriMarketing 35*(9), 34–37.

National Agricultural Statistics Service (NASS). (1997). *Farm computer usage and ownership*. Released in July 1997 by NASS, U.S. Department of Agriculture, Washington, D.C.

National Agri-Marketing Association. (1998). *1998 survey of marketing and communications expenditures by U.S. agribusinesses*. St. Louis, Mo.

National Project in Agricultural Communications. (1960). *The first seven years*. East Lansing: Michigan State University.

"National Research Report Indicates Major Shift in Farmer Information Needs." (1996). *AgriMarketing 34*(10), I6–I8.

O'Malley, S. (1994). "The Rural Rebound." *American Demographics 16*(5), 24–29. <http://www.demographics.com/publications/ad/94_ad/9405_ad/ad575.htm>.

Paskoff, B.M. (1990). "History and Characteristics of Agricultural Libraries and Information in the United States." *Library Trends 38*(3), 331–49.

Reiman, A.M. (1996). "DTN Releases Newest Ag Marketing Tool." *DTN FarmDayta On-Line* (Sept.) http://www.dtn.com/corp/press/1998/online.ctm.

Reisner, A. (1990). "Course Work Offered in Agricultural Communication Programs." *Journal of Applied Communications 74*(1), 18–25.

Reisner, A. (1990). "An Overview of Agricultural Communications, Programs and Curricula." *Journal of Applied Communications 74*(1), 8–17.

Richardson, J. G., and Mustian, R.D. (1994). "Delivery Methods Preferred by Targeted Extension Clientele for Receiving Specific Information." *Journal of Applied Communication 78*(1), 22–32.

Rogers, E.M. (1983). *Diffusion of innovations*. New York: Free Press.

Rogers, E.M., and Chaffee, S.H. (1983). "Communication as an Academic Discipline: A Dialogue." *Journal of Communication 33*(3), 18–30.

Rosengren, K.E. (1983). "Communication Research: One Paradigm, or Four?" *Journal of Communication 33*(3), 184–207.

Salomon, G. (1981). *Communication and education*. Beverly Hills: Sage.

Schudson, M. (1978). *Discovering the news: A social history of American newspapers*. New York: Basic Books.

Schwartz, E. (1995). *Econ 11/2*. New York: Avon.

Skinner, J.S. (1819). *American Farmer I,* 5.
"Sneak Peek at the Findings." (1998). *AgriMarketing 39*(9), 20.
Spiegler, M. (1995). "Hot Media Buy: The Farm Report." *American Demographics 17*(10), 18–19 <http://www.demographics.com/publications/ad/95_ad/9510_ad/ad811.htm>.
Tucker, M.A. (1996). "Ferment in Our Field: Viewing Agricultural Communication Research From a Social Science Perspective." *Journal of Applied Communications 80*(3), 25–41.
United States Department of Agriculture. (1998). *1998 agriculture fact book.* Washington, D.C.
Volk, J. (1997). "How Ag Advertisers Buy Print Media." *AgriMarketing 35*(7), 84.
Watkins, T.H. (1993). *The Great Depression: America in the 1930s.* New York: Little, Brown and Company.
Wik, R.M. (1988). "The USDA and the Development of Radio in Rural America." *Agricultural History 62*(2), 177–88.
Wik, R.M. (1981). "The Radio in Rural America During the 1920s." *Agricultural History 55*(4), 339–50.
Wobbe, S.O. (1998). "Total Recall: Wave 1 of Ag Media Study Indicates Farmers Have a 'good media mix.'" *AgriMarketing 36*(10), I4–I5.

Index

9 780813 821672